建筑立场系列丛书 No.31

微工作·微空间
Minor Works

中文版

韩国C3出版公社 | 编
王凤霞 杨蕙 于风军 | 译

大连理工大学出版社

4
- 004 欧罗巴城市 _ BIG
- 008 赫尔辛堡的H+城市重建项目 _ Erik Giudice Architects
- 012 Parkhill _ Nice Architects
- 016 赫尔辛堡医院的扩建 _ Schmidt Hammer Lassen Architects
- 018 耶路撒冷自然与科学博物馆 _ Schwartz Besnosoff + SO Architecture
- 024 太阳伞 _ Derek Pirozzi

66 微工作·微空间

- 066 微工作·微空间 _ Alison Killing
- 072 Stardom娱乐公司办公室的重建 _ D·Lim Architects
- 082 卡萨雷克斯办公室 _ FGMF Arquitectos
- 092 森林中的办公室 _ SUGAWARADAISUKE
- 100 Rubido Romero基金会 _ Abalo Alonso Arquitectos
- 110 马德里博坦基金会的新办公室 _ MVN Arquitectos
- 120 Kirchplatz办公室和住宅 _ Oppenheim Architecture + Design
- 136 Torus _ N Maeda Atelier
- 148 莫托萨布住宅sYms _ Kiyonobu Nakagame Architects and Associates

C3 建筑立场系列丛书 No.31

城市改造
26 公民参与

- 026 新型关系空间 _ Paula Melâneo
- 030 交易大厅 _ Robbrecht en Daem Architecten + Marie-José Van Hee
- 044 棚式建筑 _ Haworth Tompkins Architects
- 054 墨尔本艺术中心的哈姆音乐厅 _ ARM Architecture

城市改造
154 良性发展的城市建筑

- 154 良性发展的城市建筑 _ Simone Corda
- 158 NEO Bankside大楼 _ Rogers Stirk Harbour + Partners
- 168 来福士广场 _ Steven Holl Architects

4

004 Europa City _ BIG

008 H+ City Renewal Project in Helsingborg _ Erik Giudice Architects

012 Parkhill _ Nice Architects

016 Helsingborg Hospital Extension _ Schmidt Hammer Lassen Architects

018 Nature and Science Museum in Jerusalem _ Schwartz Besnosoff + SO Architecture

024 Polar Umbrella _ Derek Pirozzi

66 Minor Works

066 *Minor Works _ Alison Killing*

072 Stardom Entertainment Office Remodeling _ D·Lim Architects

082 Casa Rex Office _ FGMF Arquitectos

092 Office in Forest _ SUGAWARADAISUKE

100 Rubido Romero Foundation _ Abalo Alonso Arquitectos

110 The New Offices of the Botín Foundation in Madrid _ MVN Arquitectos

126 Kirchplatz Office and Residence _ Oppenheim Architecture + Design

136 Torus _ N Maeda Atelier

148 Motoazabu Apartment sYms _ Kiyonobu Nakagame Architects and Associates

Urban How

26 Civic Engagement

026 *New Relational Spaces _ Paula Melâneo*

030 Market Hall _ Robbrecht en Daem Architecten + Marie-José Van Hee

044 The Shed _ Haworth Tompkins Architects

054 Hamer Hall of the Arts Center Melbourne _ ARM Architecture

Urban How

154

154 *Virtuous Urban Pieces _ Simone Corda*

158 NEO Bankside _ Rogers Stirk Harbour + Partners

168 Sliced Porosity Block _ Steven Holl Architects

欧罗巴城市 _BIG

欧罗巴城市委员会宣布,丹麦BIG建筑事务所成为欧罗巴城市国际设计竞赛的获奖者。欧罗巴城是位于法国Gonesse三角区域的一座集文化活动、娱乐消遣和商品零售于一体的800 000m²的开发项目。

欧罗巴城围绕着欧洲、多元化、城市体验及其文化的主题,将建成超前规模的文化、娱乐和零售中心。场地的潜力独特:地处大巴黎地铁快线的运输枢纽,连接巴黎周围重要经济中心,是出入巴黎近郊大都会的必经大门,与欧洲第二大机场直接相连。除此之外,场地还覆有历史悠久的农业景观,这处景观成为城市与开阔地带之间的边界的标志,人们在此处可欣赏巴黎城市天际线的全景。

BIG建筑事务所提议整合城市周围商业区所有的新建设施,将拥挤的城市与开放的景观相结合,同时探索城市和场地的绿化潜能。欧罗巴城将成为一个文化和商业中心——周围城市的集汇点,为郊区环境注入真正的城市品质。

欧罗巴城的功能区是沿着一条内部环形林荫道而设计的,环路两侧汇集文化、娱乐和零售建筑。沿着环形林荫道,自行车和公共电动交通工具搭载游客各处观赏游玩。环形林荫道为游客带来不同的空间体验以及清晰的全景。游客迷失其中,又能够找到归路。

欧罗巴城的屋顶可以被看做是一处地形丰富的景观,游客可以进入其中,观看巴黎中心天际线和拉德芳斯车站的全景。屋顶四周的高度不尽相同,以形成缓坡,缓坡上覆有山谷和山峰景观。位于城中央的大型公共公园形成一条贯穿南北的绿化带,以将Carré Vert和Buttes St-Simon的历史景观连接起来。人们在高速公路和附近的街区可以看见公园,而公园作为绿化开发项目,已成为欧罗巴市的象征,并且为休闲与娱乐设施的融合提供唯一的可能性。

Europa City

The Europa City Committee announced Danish architecture firm BIG as the winner of an international competition for the design of Europa City, a 800,000 square meter cultural, recreational and retail development in trigonal field of Gonesse, France. Europa City will offer an unprecedented scale a mix of retail, culture and leisure around a defining theme: Europe, its diversity, its urban experiences and its cultures. The site has exceptional urban potentials: Locally as a main node on the Grand Paris Express Metro connecting key economic hubs around Paris, regionally as an entrance gate to the metropolitan area of Ile

区域 / area | 规划分区 / division of program | 沿街规划区 / the program along the street | 规划成环形 / looping the program | 优化规划 / optimized the program | 适应场地 / site adaptation

de France and internationally with its direct connection to the second largest airport in Europe. Additionally the site is occupied by a historic agricultural landscape marking the edge between city and open land and offers a panoramic view to the skyline of Paris.

BIG propose to integrate the new facility in the surrounding business district as an urban form that combines dense city with open landscape, exploring the urban and green potentials of the site at once. Europa City will become a cultural and commercial center – a gathering point for the surrounding cities, injecting true urban qualities into the suburban environment.

The programs of Europa City are organized along an internal circular avenue with a mix of retail, entertainment and cultural programs on both sides. Along the avenue bicycles and electric public transport bring visitors quickly around, the circular avenue creates a variety of spatial experiences and a clear overview – It allows you to get lost,

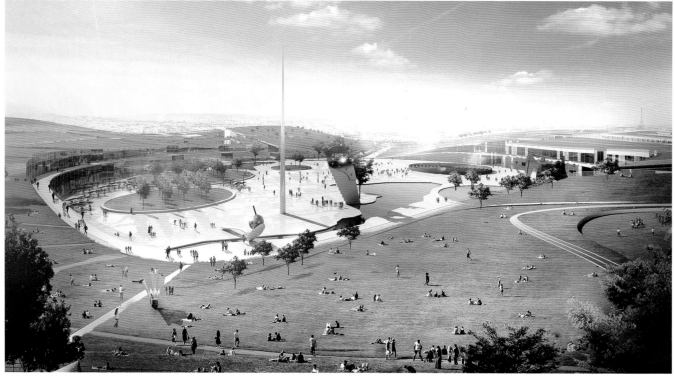

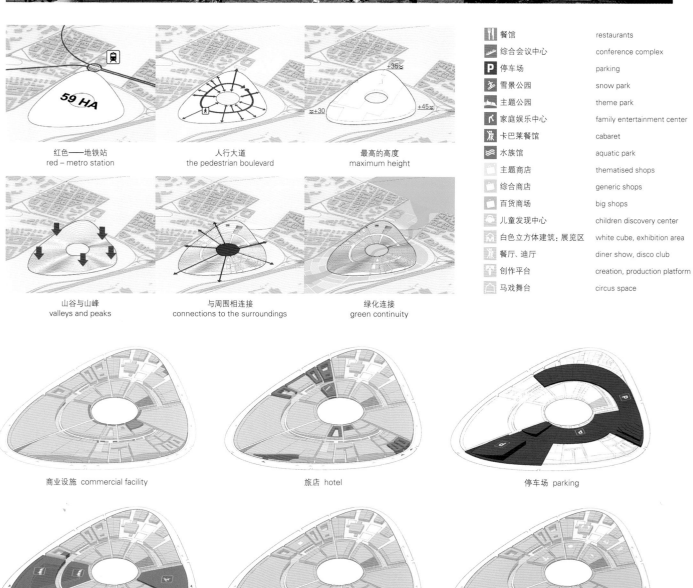

餐馆	restaurants	
综合会议中心	conference complex	
停车场	parking	
雪景公园	snow park	
主题公园	theme park	
家庭娱乐中心	family entertainment center	
卡巴莱餐馆	cabaret	
水族馆	aquatic park	
主题商店	thematised shops	
综合商店	generic shops	
百货商场	big shops	
儿童发现中心	children discovery center	
白色立方体建筑：展览区	white cube, exhibition area	
餐厅、迪厅	diner show, disco club	
创作平台	creation, production platform	
马戏舞台	circus space	

红色——地铁站 red – metro station
人行大道 the pedestrian boulevard
最高的高度 maximum height
山谷与山峰 valleys and peaks
与周围相连接 connections to the surroundings
绿化连接 green continuity

商业设施 commercial facility
旅店 hotel
停车场 parking
公园 park
餐馆 restaurant
休闲设施 leisure facility

and still find your way.

The roof of Europa City is conceived as an accessible topographic landscape, allowing visitors to experience the panoramic views to the skyline of central Paris and La Defense. The height is varied along the perimeter, forming a gently sloping landscape of valleys and peaks. In the center a large public park forms a north-south green connection between the historic landscapes of Carré Vert and the Buttes St-Simon. Visible from both highway and urban neighborhoods, it becomes symbolic of Europa City as green development, and offers unique possibilities for integration of recreation and entertainment facilities.

项目名称：Europa City
地点：Paris, France
建筑师：Bjarke Ingels, Andreas Klok Pedersen
项目领导：Joao Albuquerque, Gabrielle Nadeau
项目团队：Maren Allen, David Tao, Salvador Palanca, Marcos Bano, Lucian Racovitan, Ryohei Koike, Camille Crepin, Elisa Wienecke, Lena Rigal, Paolo Venturella, Tiina Liisa Juuti, Jeff Mikolajewski
合作商：Tess, Transsolar PR, Base, Transitec, Michel Forgue
甲方：Groupe Auchan
用地面积：800,000m²
设计时间：2012

赫尔辛堡的H+城市重建项目 _Erik Giudice Architects

2012城市再生WAN奖宣布Erik Giudice建筑师事务所为赫尔辛堡的H+城市重建项目的获奖者，该事务所因其对瑞典赫尔辛堡城市重建提出的创意解决方案而获此殊荣。WAN评委成员如是说，"这一方案营造一种鼓励居民相互接触交往的气氛和能力，使以运河为主载的中央轴向城市重新连接起来，并且创造了一系列服务于运河的公共空间，且这些空间还与场地的关键性运动相关联。

蓝色带与绿色带相衔接

H+项目是瑞典最引人瞩目的城市重建计划之一。这个正在实施的规划旨在通过"蓝色带与绿色带相衔接"，来对赫尔辛堡市的南部进行彻底的改造，使其与大海连接起来，成为一处富有特色的水景。

项目区域位于老城区和海湾中间，即第一处进行改造的区域，既是整个H+再生项目的"试验田"，也在H+城市项目中保留自己的特色。

赫尔辛堡大学校园和若干活力四射的公司位于此地，但这一地区匮乏住宅、公共设施、活动场所。改造的目的是将这一区域改建为综合性的城区，继续保持企业的精神文化，并且促进企业和大学间的合作交流。

步行街的长度和宽度不一，令人耳目一新。建筑的规模和类型各异，新旧融为一体，通过结合一系列联系紧密且宽敞的公共空间，来形成具有活力的城市肌理。

"运河"沿岸的房间宽度不尽相同，且深浅不一，新颖独特。建筑规模各不相同，以形成一系列联系紧密且更加宽敞的公共空间。

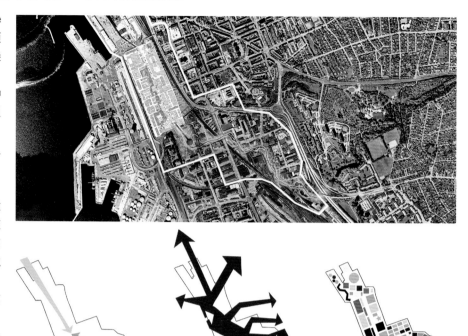

blue meets green　　integration　　diversity

沿运河所在的地面层而行，建筑一层是教育机构、咖啡馆、餐馆和办公区域。多功能建筑群的形式成为一种催化剂，非常具有战略性，可以进一步拓展运河的空间。现存建筑的楼层平面面向公众开放，且围以玻璃，以创造全新的联系。住宅区面向运河沿岸的房间，而这些房间可以用于举办更加丰富的活动。

街区紧邻西部，以免受风沙、噪音以及危险品带来的危害。开放的庭院直通运河，与河水相连，且可以观赏到运河的美景。办公区面朝基础设施一带，而较高的居民区则面对中央运河和绿化带。

蓝色带与绿色带相衔接的区域被一处纵向的知识汇集带所强化。沿着中央运河设置学校和幼儿园，以形成一个知识带，且与大学毗邻。其他形式的培训、成人教育以及商业课程机构也主要规划在运河沿线上，以构成这一地区的主要特色，所有的教育机构实际上是连为一体的。

H+ City Renewal Project in Helsingborg

Erik Giudice Architects were recently announced as the H+ Bredgatan Winner of the WAN Awards 2012 Urban Regeneration for their innovative solution in Helsingborg, Sweden. "The social atmosphere and ability to encourage residents' integration reconnect the city with a central spine around a canal theme, creating a number of public spaces that are servicing the scheme but also connecting with the key movement around the site", according to WAN jury members.

Blue-green Connection

The H+ project is one of Sweden's most ambitious planning and urban renewal projects. The ongoing process aims to radically transform the southern parts of Helsingborg connecting them to the sea through the "Blue-green connection", a landscaped water feature.

This area is strategically located between the old city and the harbor, and will be one of the first areas to undergo transformation, it not only will serve as a "test-bed" for H+ at large, but will also be given its own identity in the H+ urban mosaic.

The area is already hosting the Helsingborg University Campus, and several dynamic companies, but lacks of housing, public services and has a poor public space. The aim is to transform the area into a mixed urban fabric, keeping the spirit of entrepreneurship and enhancing the collaboration between university and companies.

The varying width and depth of the central

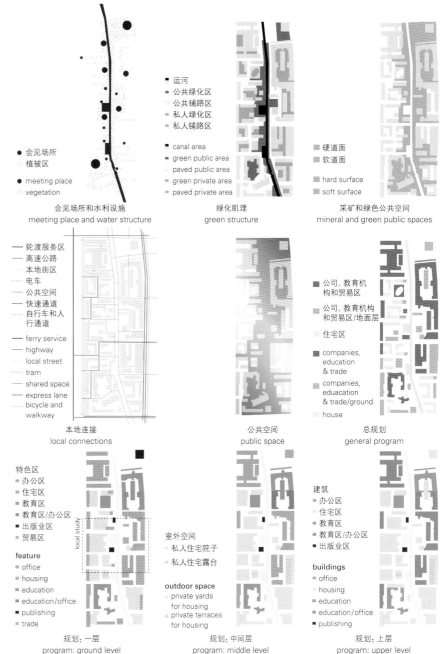

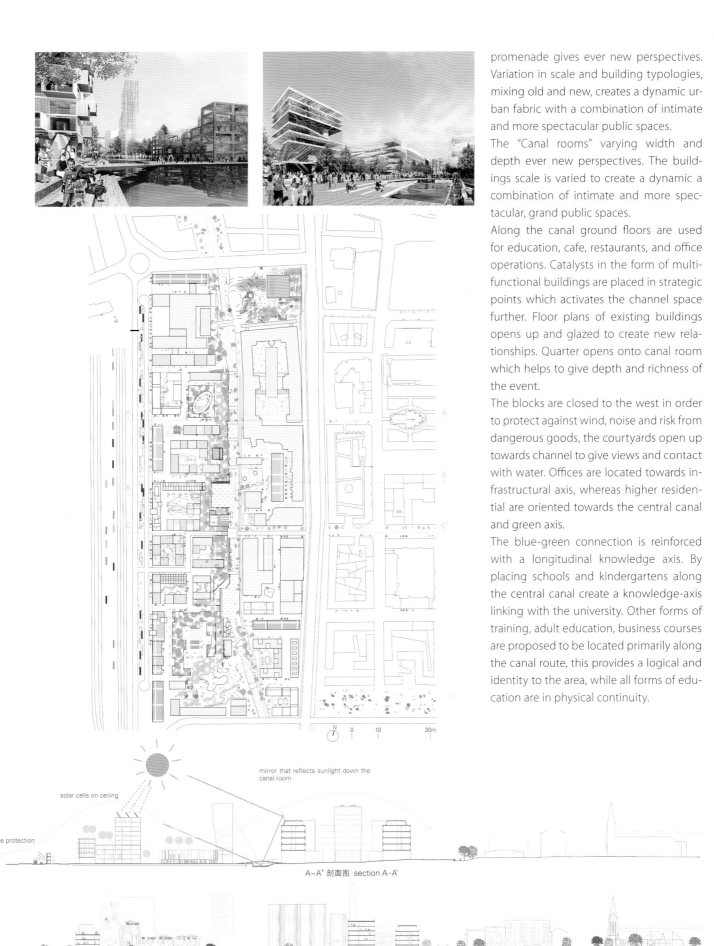

promenade gives ever new perspectives. Variation in scale and building typologies, mixing old and new, creates a dynamic urban fabric with a combination of intimate and more spectacular public spaces.

The "Canal rooms" varying width and depth ever new perspectives. The buildings scale is varied to create a dynamic a combination of intimate and more spectacular, grand public spaces.

Along the canal ground floors are used for education, cafe, restaurants, and office operations. Catalysts in the form of multi-functional buildings are placed in strategic points which activates the channel space further. Floor plans of existing buildings opens up and glazed to create new relationships. Quarter opens onto canal room which helps to give depth and richness of the event.

The blocks are closed to the west in order to protect against wind, noise and risk from dangerous goods, the courtyards open up towards channel to give views and contact with water. Offices are located towards infrastructural axis, whereas higher residential are oriented towards the central canal and green axis.

The blue-green connection is reinforced with a longitudinal knowledge axis. By placing schools and kindergartens along the central canal create a knowledge-axis linking with the university. Other forms of training, adult education, business courses are proposed to be located primarily along the canal route, this provides a logical and identity to the area, while all forms of education are in physical continuity.

Parkhill _Nice Architects

想象一下，一处古老的室外露台剧场，实际上并不适合住宅开发，地势过于陡峭，导致交通不便，北坡无日照，使其无法适当地采光——这些都是这一场地的主要特征。然而，这里却毗邻布拉迪斯拉发最古老的公园，并且人们还可以一览布拉迪斯拉发最壮观的风景。

近期，这片52 000m²的区域成为斯洛伐克共和国城市设计与建筑竞赛的舞台。建筑师们利用地势较低区域设置的公共服务区，来开发高密度的住宅区。竞赛要求之一是保留原有的V形栗子园，而其余部分作进一步的开发。

斯洛伐克建筑公司Nice建筑师事务所赢得了2012 Parkhill城市建筑竞赛的大奖，以开发人类居住的居民区与工作区共生的社区。Nice建筑师事务所在场地四周设置了规划好的体量，有效解决了与其他区域相隔离的难题，并且为沿场地边缘修建的地下车库提供了机会，同时将中心地带壮观的公共绿化空间与毗邻的栗子园连接起来。建筑师设计了一座小型步行公园，人们在公园里可以进行各种各样的户外活动，有效地对这一区域进行了一定的控制，为孩子们提供了一处安全的嬉戏场所。

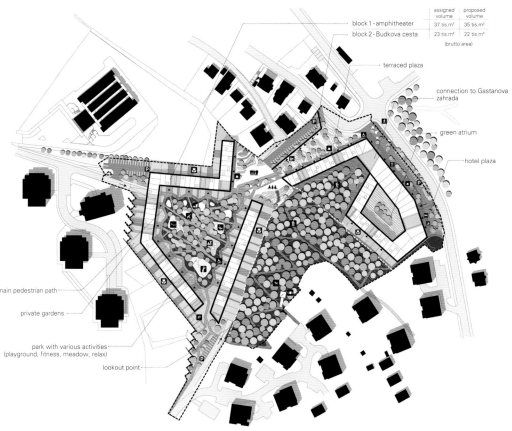

这处住宅实质为联排住宅,其楼层平面设计较为灵活,立面灵活多变。其灵感来自于附近街区30步或50步距离的现代别墅,这一理念建造了符合人类尺度、紧凑且具有各式各样公共空间的建筑。

建筑的构成也不禁让人想起曾经的圆形剧场。

Parkhill

Imagine former site of old outdoor amphitheater virtually unsuitable for residential development. Terrain too steep for comfortable traffic, sloping northward preventing proper insolation – these are the main characteristics of the plot. On the other hand, it is offering spectacular views on Bratislava and proximity of one of the oldest park in Bratislava.

This 52,000m² site happened to be a stage of most recent urban design and architectural competition in Slovak Republic. Assignment was to create intensive resi-

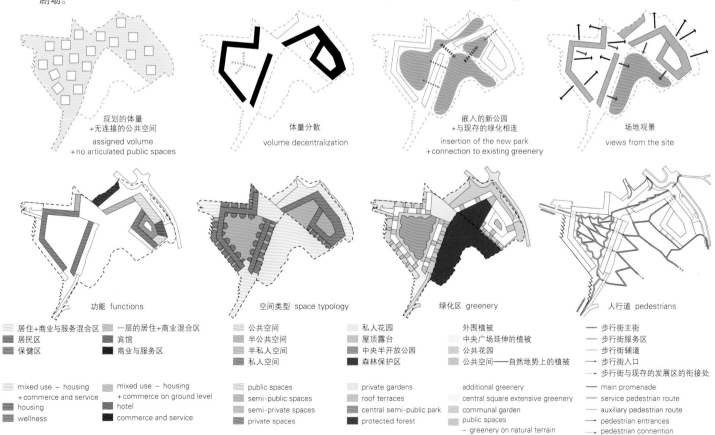

dential development with public services located in the lower part of the site. One of the requirements was to preserve V-shaped chestnut garden leaving the rest of the site for further development.

The Slovakia architectural firm Nice Architects won the Parkhill urbanistic architectural competition launched in 2012 to develop living and mixed used block. Nice Architects distributed proposed volume along the edges of the site. It helped to solve insolation problem and gave the architects an opportunity to place underground garages along the edge of the site leaving the heart of the zone for grandiose and green public space connected with adjacent chestnut garden. They created a small park without traffic which offers variety of outdoor activities and provides a certain control for the area and safety for the children to play.

Residential houses were designed essentially as row houses with flexible floor plan and set of various facades. Inspiration for them can be traced to modern villas from 30-ties and 50-ties present in the neighborhood. This idea helped to create architecture characterized by human scale, local connection and variety of public spaces.

Final composition of the buildings is also serving as small reminiscence of the former amphitheater.

项目名称: Parkhill
地点: Budkova cesta, Bratislava, Slovakia
建筑师: Nice Architects
项目团队: Tomáš Žáček, Soňa Pohlová, Igor Žáček
景观建筑师: 2ka Landscape Architects
竞赛名称: Parkhill competition winning proposal
设计时间: 2012

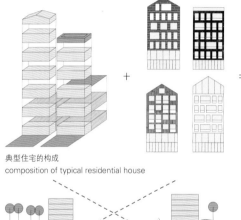

典型住宅的构成
composition of typical residential house

基本模块(不包括车库)
basic module (without garage)

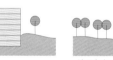

所需的停车场
required parking

地下车库+步行庭院
underground garages+trafficless courtyard

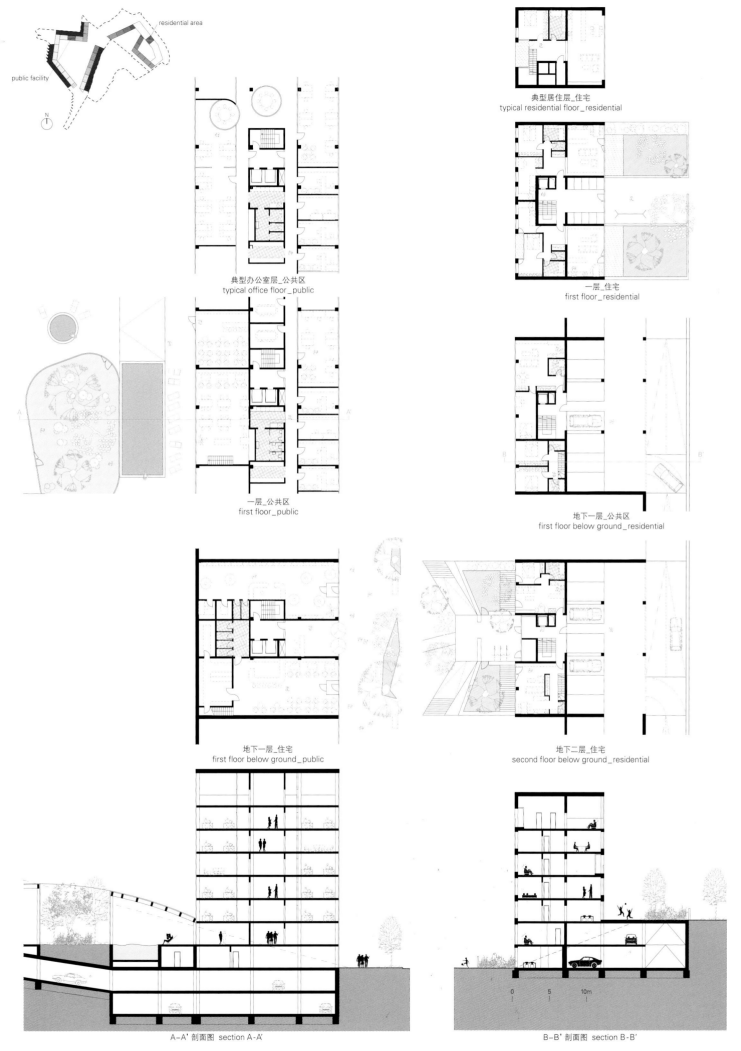

与城市相连接 URBAN CONCERNED

赫尔辛堡医院的扩建 _Schmidt Hammer Lassen Architects

Schmidt Hammer Lassen建筑师事务所在斯科纳地区举办的设计竞赛中获胜,将对瑞典南部的赫尔辛堡医院进行扩建。这个项目面积为35 000m²,包括为成人精神病患者的新病区、门诊部以及医疗实验室等。整个设计方案的关键是灵活性、清晰的布局、多样性、人体尺度、绿化庭院以及最佳日光条件。

Schmidt Hammer Lassen建筑师事务所的合伙人Kasper Frandsen表示:"我们设计了一座单体建筑,使其与现有医院及其周围城区联系起来。"他还说:"这座建筑以雕塑的形式呈现出来,且容纳了三处活动区,低层是门诊部和实验室,上层为精神病人病房(其结构更加透明)。"

整栋建筑设计灵活,将来用途有所改变,或者是有功能上的调整也十分简便。不断变化的锯齿状立面能够根据功能的需要,创造不同的空间,并使它们适应各种带有开放或封闭部分的结构。走廊构成了建筑的脊梁,以清晰的方式把整个空间不同功能组合起来。它既是一条充满活力的城市街道,也是广场、绿色庭院的视野相互交叉的一个网络。

在精神病房,设计的重点放在为病人提供放松的和活动空间的环境方面,这能够让病人感到静下心来,受到鼓舞。病区的布局非常清晰,绿色的屋顶景观独特,且与外界相隔离。病房还带有一个覆顶的内部小庭院,安静,给人以安全感。在上层,病人能够欣赏城市和厄勒海峡全景,同时屋内也能享受到充沛的阳光。

整栋建筑的功能布局尽量减少奔波之苦,优化了日常操作流程,对不同区域的活动需求进行调整。

"整座医院的设计理念的重点是以人为本。虽然建筑师知道医院必须要有合理高效的诊疗功能,但是建筑师从不会忘记这些都是以人为基础的——每一个有灵魂、有感知和有理解能力的个体",Schmidt Hammer Lassen建筑师事务所的合伙人Kim Holst Jensen说道。

Helsingborg Hospital Extension

Schmidt Hammer Lassen Architects has won the competition organized by Region Skåne, to design the 35,000 square meter extension to the Helsingborg Hospital in the southern part of Sweden. The project comprises a new ward for adult psychiatry, an out-patient clinic and medical laboratories. Keys to the whole design have been flexibility, a clear layout, variety, human scale, green courtyards and optimal conditions for daylight.

"We have designed a single building, whose architecture relates both to the existing hospital and the surrounding city," explained Kasper Frandsen, Associate Partner at Schmidt Hammer Lassen Architects.

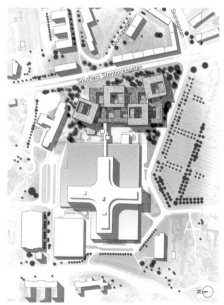

"The building is expressed in one sculptural form, which houses three areas of activity: the out-patient clinic and laboratories in the lower levels of the building, while the top levels containing the psychiatric ward open up to a more transparent structure." The building is flexible and therefore sustainable with regards to future demands for changing use and functions. The shifting and indented facade creates varying spaces and makes it possible to adapt the structure with open and closed parts depending on the functions behind it.

项目名称：Extension of Helsingborg Hospital
地点：Helsingborg, Sweden
建筑团队：Schmidt Hammer Lassen Architects, Aarhus Arkitekterne, NNE Pharmaplan
景观建筑师：Kragh & Berglund Stockholm
甲方：Region Skåne
用地面积：26,000m²
有效楼层面积：35,000m²
建筑规模：one story below ground, 4 stories above ground
获奖情况：2013, 1st prize proposal

A hallway makes up the spine of the building and gathers together the different functions in a clear fashion. It has the double function of a dynamic urban street with a fine net of intersections, squares and views to green courtyards.

In the psychiatric ward, the emphasis is on an environment that allows for both relaxation and activities. This will have both a calming and inspiring effect on the patients, who are thereby challenged in a secure setting. The layout of the ward is clear, and the green roofs establish a distinct, undisturbed landscape. The composition of the bed wards creates sheltered, inner courtyards signaling calm and safety. From the upper levels, the patients have a panoramic view over the city and Oresund, which in turn affords plenty of daylight in the rooms.

The layout of the main functions of the building minimizes walking distances, optimizes daily operations and adjusts to the special needs of the different areas of activity.

"Human beings are the focal point of our approach to designing hospitals. Though we know that the buildings must support rational and efficient clinical operations, we never forget that this is all about people – about individuals with a soul and ability to feel and understand," observed partner at Schmidt Hammer Lassen Architects, Kim Holst Jensen.

融入环境 CONTEXTUAL INTEGRATION

耶路撒冷自然与科学博物馆 _Schwartz Besnosoff+SO Architecture

耶路撒冷自然与科学博物馆的公共设计竞赛于2012年举行。博物馆位置显著，位于耶路撒冷市中心，毗邻以色列政府的议会大楼。Schwartz Besnosoff+SO建筑事务所最终获得了竞赛的胜利，并规划了一座开放的、包容的、令人震撼的建筑。

建筑的设计灵感来源于设计师高度的城市与环境共生意识。这是任何现代建筑所必需的，尤其是在着重于体现人与自然复杂而又多方位的关系的场地中。

建筑计划融入现存的空间模式中，在遵循现有的规则同时，使空间更加开放，引导人们进入一座带有地下室的建筑，而这座建筑则会引导人们穿过公园，进入一排房屋建成的博物馆中。建筑师没有强化其商业性特点，而是赋予其绿色的特性，以吸引和邀请人们来到博物馆、天文馆以及停车场的入口处。建筑师相信通过在内部设置公园的方式，对场地的增建工程将会在公园与以色列博物馆之间创造一条连续的绿化带。这条绿化带穿过自然博物馆的场地，越过对面的即将建成的国家图书馆和财政部，到达以色列国会大厦对面的玫瑰园。一个意义重大的建筑体系由此将创建——一处较为民主的空间，邀请城市的居民参加，并且成为一处以色列的代表性空间。在建筑师看来，一座拉近展厅和展品距离的博物馆也可拉近公民和民选官员之间的沟通距离。

建筑师精心考虑了博物馆内的参观路线，并将其作为核心原则。博物馆各展厅之间的参观路线为线形，这些路线均位于变化的展厅系统之间，有些展厅呈倾斜状态，有些则是位于平地。整体路线可以被规划为一个带有中轴的线形路线，没有不同的岔道和路线选择。在标高为-12.00m的位置，参观者乘电梯直达博物馆的顶层，到达室外展厅，人们从屋顶展厅下至天体物理展厅，再到出口/入口。这种路线规划创造了一种场景，即人们在屋顶的移动路线是博物馆路径的一部分，建筑师认为这种体验非常重要。

Nature and Science Museum in Jerusalem

Public competition for Planning the Natural and Science Museum in Jerusalem was held on 2012. The noticed location was next to Israel's government assembly building, the heart of Jerusalem. The winning proposal was the plan of Schwartz Besnosoff and SO Architecture, which was an open, absorbent, breathing building. The planning of the building was inspired by a high level of urban and environmental awareness, required in any contemporary spatial project, and all the more so in a place intended to highlight the complex and multifaceted relationship between humans and nature.

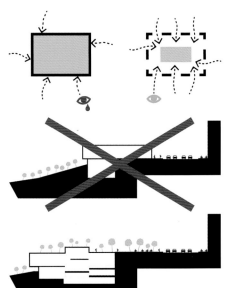

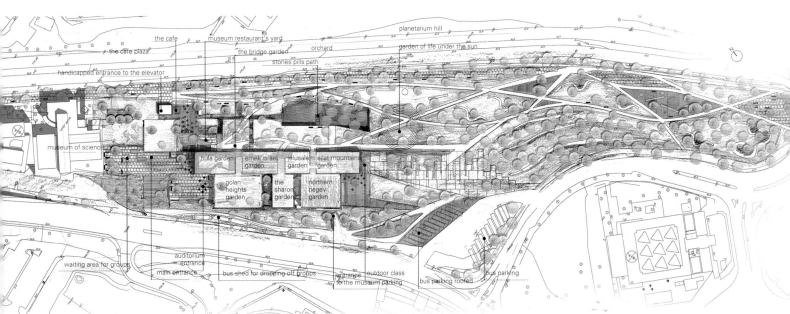

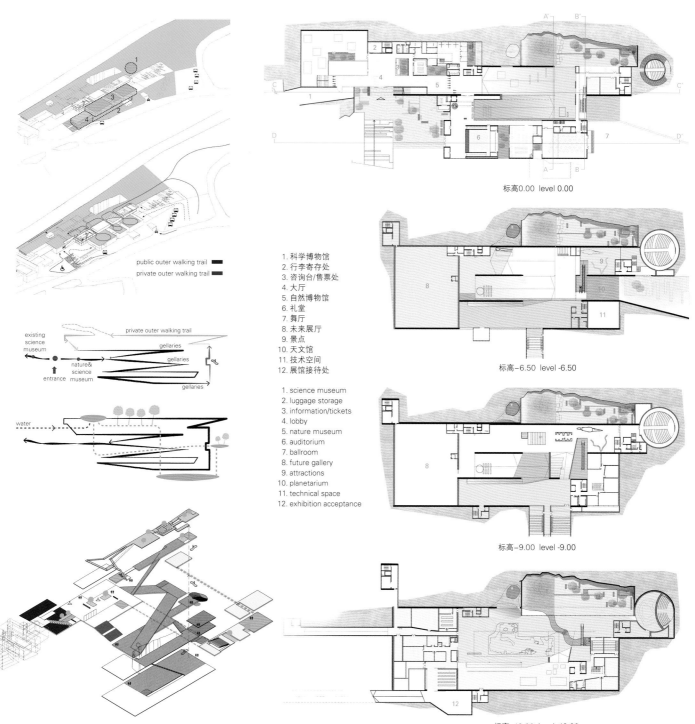

summer public open space, ventilated

winter public open space, protected against winds

The structure's geometry blocks the unwanted wind, the wanted wind can be put in "fingers" for ventilation in the summer.

summer lots ventilated by western winds

winter lots protected from the western wind in the winter

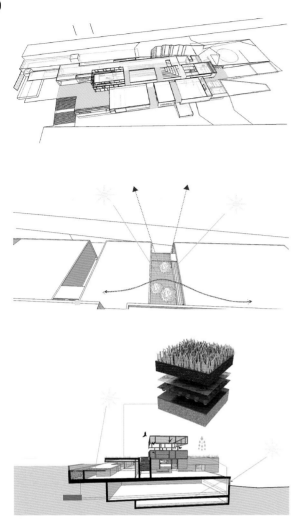

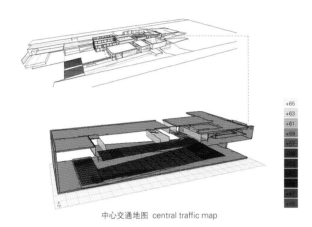

中心交通地图 central traffic map

The decision to integrate into the patterns of the existing space and follow its rules while empowering the open space led the architects to the choice of a basement building that leads to the row of museums through the park. Instead of a forced commercial character, the architects propose a green character that draws and invites people to the entrances to the museum, planetarium, and parking lot. They believe that intervention in the site by means of a built-in park will create a contiguous green space from the park to the Israel Museum, through the Nature Museum site, passing between the National Library to be built opposite it and the Ministry of Finance, to the Rose Garden across from the Knesset building. Thus a meaningful system will be created – a democratic space that invites the citizens of the city and the state to the representative space of Israel. In the architects' view, a building that negotiates between field and object may promote spatial negotiation between the citizens and their elected officials.

The architects consider the movement within the museum as the core principle. The course of the museum movement is characterized as a linear path between fields, within a system of meta-galleries, some inclined and some flat. The movement was organized as a central, linear system, with different interactions and options. Upon arriving at Level –12.00, an escalator takes the visitors up to the roof of the museum – to the outside exhibition gallery. From the roof exhibitions one descends to the astrophysics gallery and from there to the exit/entrance. The movement plan creates a situation in which the movement on the roof is part of the formal museum path; it was considered very important.

项目名称：Nature and Science Museum in Jerusalem
地点：Jerusalem, Israel
建筑师：Gaby Schwartz, Shachar Lulav, Oded Rozenkier
项目团队：Gaby Schwartz, Shachar Lulav, Oded Rozenkier, Alejandro Fajnerman, Levav Shachar, Tomer Nachshon, Noa Hefez, Noy Lazarovic, Omry Schwartz, Boaz Rotem
组织者：Jerusalem Municipality, The Jerusalem Development Authority, The Hebrew University of Jerusalem, The Jerusalem Foundation and Bloomfield Science Museum in cooperation with the competition committee of Israel Association of United Architects
花园面积：7,000m²
用地面积：30,000m²
有效楼层面积：23,000m²
建筑规模：five stories below ground, four stories above ground
设计时间：2012.7

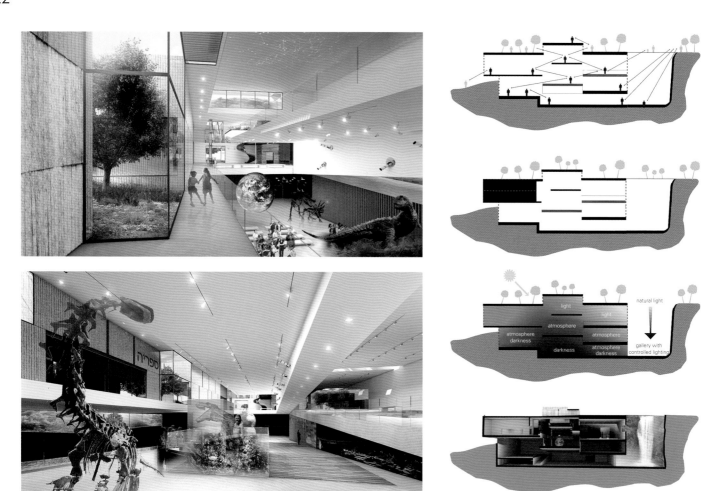

规划 program

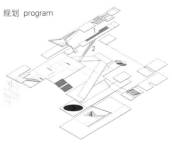

规划1 program 1

规划2 program 2

规划3 program 3

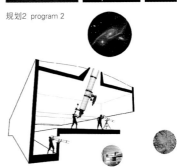

规划4 program 4

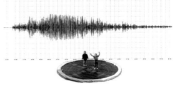

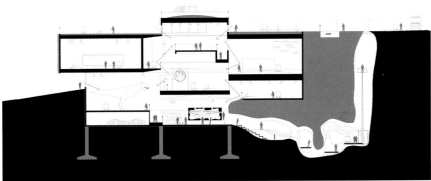

A-A' 剖面图 section A-A'

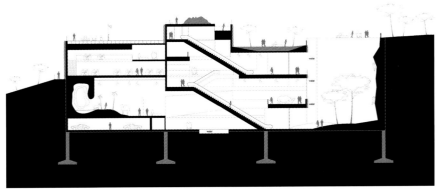

B-B' 剖面图 section B-B'

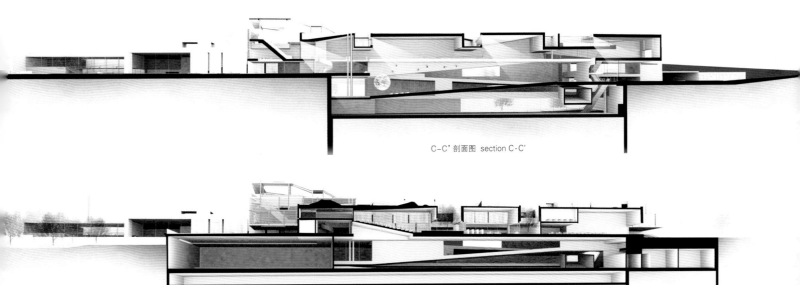

C-C' 剖面图 section C-C'

D-D' 剖面图 section D-D'

太阳伞 _Derek Pirozzi

2013年摩天楼设计竞赛由eVolo建筑事务所和设计杂志发起，其获奖作品是Derek Pirozzi的太阳伞项目。这是一个充满活力的摩天楼规划，它通过减少地层表面的热增量以及冰冻海洋水的方法，来重建北极冰帽。

在全球变暖的后几十年里，极地冰帽的温度急剧上升，使得南极和北极的冰架开始变薄，有断裂的痕迹，并且开始融化，流进海洋中。

这个规划的首要目标是重建北极冰层，它能够在脆弱的北极地区通过减少热增量，来降低地层表面的温度。

太阳伞的上层结构展示了人们对受保护的北极区在未来即将资源枯竭时所采取的防御措施。由于它拥有淡化海水和发电的功能，这座位于北极的摩天楼还被美国国家海洋和大气局研究实验室装备为一座漂浮在空中的城市，以及具有可再生能源的发电站。这个项目的住宅单元风格类似于宿舍，既是一处生态旅游景点，也是野生动物的生态栖息地。这一系列的结构都将被战略性地安置在受影响最大的区域。

盐水通过一种位于建筑核心区的渗透压（海水盐差发电）来产生可再生的能源。此外，该结构带有面积较大的天篷，在北极地表面减少热增量的同时，也能获取太阳能。值得关注的是，这座太阳伞的保温表皮拥有一系列的模块，这些模块是由用于抽取盐水的聚乙烯管道系统构成。最后，太阳伞也能通过使用收集室（用来冰冻海水）来对冰帽进行重建。

Polar Umbrella

The winner of the 2013 Skyscraper Competition, organized by eVolo architecture and design magazine, was awarded to Derek Pirozzi for his project, Polar Umbrella. The proposal is a buoyant skyscraper that rebuilds the arctic ice caps by reducing the surface's heat gain and freezing ocean water.

During the last decades of global warming, the polar ice caps have experienced a severe rise in temperature causing the north-

项目名称：Polar Umbrella
地点：the North Pole, the South Pole
建筑师：Derek Pirozzi
竞赛：2013 Skyscraper Competition

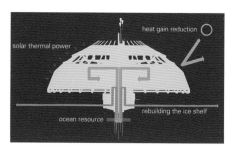

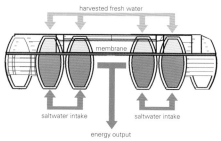

ern and southern ice shelves to become thin, fractured, and melt into the ocean. Rebuilding the arctic layers is the primary objective of this proposal which cools down the Earth's surface by reducing heat gain in vulnerable arctic regions.

The Polar Umbrella's super-structure becomes a statement for the prevention of future depletion of People's protective arctic region. Through its desalinization and power facilities, this arctic skyscraper becomes a floating metropolis equipped with NOAA (National Oceanic and Atmospheric Administration) research laboratories, renewable power stations, dormitory-style housing units, eco-tourist attractions, and ecological habitats for wildlife. A series of these structures would be strategically located in the most affected areas.

Salt water is used to produce a renewable source of energy through an osmotic (salinity gradient power) power facility housed within the building's core. In addition, the structure's immense canopy allows for the reduction of heat gain on the arctic surface while harvesting solar energy. The umbrella's thermal skin boasts a series of modules that are composed of a polyethylene piping system that pumps brackish water. Finally, the Polar Umbrella also regenerates the ice caps using harvest chambers that freeze the ocean water.

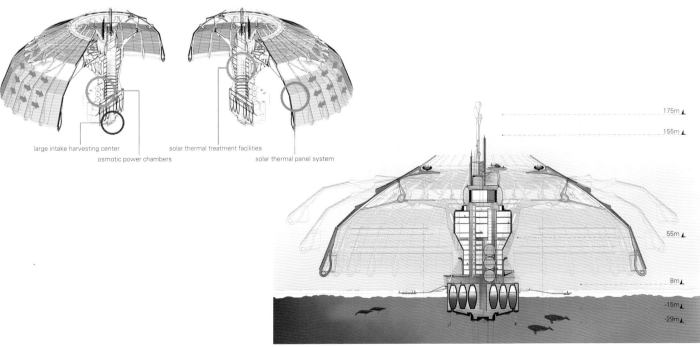

公民参与
Civic Engagement

　　传统的城市公共空间模式——街道、广场或者是林荫大道——在其使用方面不再能满足人们的要求，同时也变得完全过时。规划者们不能继续认为公共空间还处于象征、类型和形态的逻辑内。在现代城市社会中，本地与全球之间的差距正在逐渐缩小，且通过全新的交流方法和技术（正在改造城市生活和空间适应性）来减小。因此，从公共空间的角度来说，当代城市要求对空间进行翻新，而从建筑的角度来看，则要求建造一处完全不同的充满活力的场所——一种需要成为新型城市关系的新办法出现在人们的面前。

　　这导致我们要询问公共空间在今天应该具有哪些功能。它作为一种长期可用的活动空间，怎样对新需求和使用形式做出回应？城市规划者和建筑师们对此应该设计出什么样的战略，以对当代城市内的新空间与现存的公共空间进行布局？

　　以下所介绍的项目或包含在一个大型的整体规划内，或是一个独立的个体，但是在公共空间方面均发挥着重要的作用。它们或成为一个参考点，或是一处地标，定义且激活了新用途，与周边环境和谐共处，并且对市民和推动者或组织之间的距离进行了调和。

Traditional models for cities' public spaces – the street, the square or the boulevard – are no longer enough or are completely obsolete in their use. Planners cannot continue to think the public space within the symbolic, typological and morphological logic. In contemporary urban societies differences between local and global are fading away, fostered by the new communication methods and technologies that are transforming urban life and space appropriation. Consequently, contemporary urban public demands from public spaces a renovated mobility and from buildings a different dynamic – new answers needed as new citizenship relations are emerging.

This leads us to ask which functions should public space assume today and how can it respond to new needs and patterns of use, as an always-available space for action? And how can urban planners and architects create design strategies for that, tracing configurations for new and existing public spaces, in the contemporary city?

Included in larger masterplans or acting individually, the analyzed buildings have an important role in the perception of the public space. They can be references or landmarks, define or activate new uses, harmonize relations with the surroundings or mediate the proximity between the population and the promoters or organizations.

交易大厅
/ Robbrecht en Daem Architecten + Marie-José Van Hee
棚式建筑 / Haworth Tompkins Architects
墨尔本艺术中心的哈姆音乐厅 / ARM Architecture
新型关系空间 Spaces / Paula Melâneo

如果说希腊集会可以被视为城市公共空间的一个起源，作为一处可以自由交换政治意见的本地场所的话，那么我们就不能忘记几个世纪之后，公共空间已经被规划成一处展示权力的地点，建有各种强加建设的公共性或者是机构性的建筑和纪念物。在中世纪时代，人们能够在街上看到国王、皇家法院和军事游行，以展示他们的权力，公共广场是教会用来处决非信徒的舞台。近些年来，许多城市都是在严格的政治理念下被构思出来的，这些政治理念具有严格的设计规则，如在苏联许多城市的大规模例子中，公共空间始于当代理想的社会主义社会的反映。

在当代社会，公共空间的使用成为一个横向反映社会的问题。可能产生的公共空间的定义带来了更多的学科间的讨论，不仅仅是建筑师、城市规划师或者是景观设计师，甚至是社会学家、人类学家、心理学家和艺术家也参与其中。从理性的层面来说，决策者和政治家应该与市民和专家共同合作，共同参与到行动中来，理解深层的问题，不仅仅要解决大众问题，还要对特殊要求进行处理。这样，引自Manuel Gauza的话说，时至今日，原公共空间才能被称之为是集体空间或者是关系空间。¹

今天，公共空间可以被视为快速交通领域或着文化思考缓慢进行的场所，同时也是举办活动的场所。最重要的，公共空间将一种混合的社会意识和民主交流融合起来。但是，如同社会学家Zydmunt Bauman写到，不同居民区的人们汇集到这种城市空间，面对面，大家偶然地遇见，和对方交流，或者向对方发动挑战，谈话、吵架、讨论或者认同对方观点。将私人问题上升至公共问题的层面上的这种事件在规模和数量上都在减少。²

因为这些空间的数量在减少，那么当代消费社会便会出现两种紧急情况。一是私有制空间，如机场和购物区，经常被用作公共空间。其中，私有发起人发挥了重要的作用，他们曾经隶属于机构（政治、经济或者是宗教方面的机构）。另外一种情况便是一种新型的社交虚拟空间正在兴起，这使我们质疑在当今时代，一处集会空间究竟应该是什么样子的。

然而，主要的城市公共空间在今天仍然是政治和公民的象征，作为潜在的公众视野所及的场地。在这里，公众可以顷刻表达他们普遍的不满或分歧。这正在伊斯坦布尔（土耳其）的塔克西姆广场发生着，在经济危机时，欧洲的许多国家也经历过。在"阿拉伯之春"期间，阿拉伯国家也发生过此类事情。公共广场成为展示对政治决策和体系不满的舞台。

对于现在的公共空间的概念来说，一个全新的关键方法是十分必要的，即在媒体的、流行的和典型的模式之间寻求一种平衡。

城市公共空间应该考虑诸如全球化、流动性、旅游流的增加或者是经济危机等事实，且能够处理好与利益相关者的关系，包括全民、机构、推动者、协会和个人。当代的城市居民关系要求公共空间具有灵活性，建筑扮演着全新的公民角色。这样便能使公众参与进来，且能够举办全新的公民活动。而这些空间能被看做是新型集会和关系空间。

Market Hall
/ Robbrecht en Daem Architecten + Marie-José Van Hee
The Shed / Haworth Tompkins Architects
Hamer Hall of the Arts Center Melbourne / ARM Architecture
New Relational Spaces / Paula Melâneo

If the Greek agora could be seen as an origin of the urban public space, as a local where political opinions were freely exchanged, we cannot forget that for several centuries after, the public space has been programmed as a place for the display of power, with imposing public or institutional buildings and monuments. In medieval times the streets saw kings, royal courts and military parades demonstrating their power, and public squares were stages for the church to execute the nonbelievers. More recently, many cities are conceived under political ideas that imposed strict rules of design, like the large scale example in USSR's cities, where public space started to be a reflection of the contemporary socialist societies ideals.

In contemporary societies, the uses of public space became an issue transversal to society. The possible definition of public space brings more disciplines to the discussion, not just architects, urban planners or landscape designers, but also sociologists, anthropologists, psychologists or artists, are among others. Ideally, decision makers and politicians should work with citizens and specialists, in participative actions, to understand deep problems and to solve not just general issues, but the specific needs. Thus, the formerly public space can now be called collective or relational space, using Manuel Gauza's words.¹

Today, public space can be seen as fast commuting areas or spots for slow cultural contemplation, but also for actions to take place. Most important is the integration of a social mixing sense and democratic exchange. However, as the sociologist Zygmunt Bauman wrote, *"such urban spaces where the occupants of different residential areas could meet face-to-face, engage in casual encounters, accost and challenge one another, talk, quarrel, argue or agree, lifting their private problems to the level of public issues are fast shrinking in size and number."* ²

As those spaces are decreasing, two emergent phenomena can be appointed in the contemporary consumer society. One is that private property spaces – such as airports or shopping areas – are often used as public spaces. This gives a significant role to private promoters, which once belong to institutions (political, economic or religious). Another is that a new virtual space for communication is rising, it makes us question what could be a gathering space nowadays.

However, main urban public spaces are still today's political and civic symbols, as potential sites of public viewing. There, the population can demonstrate immediately their general discontent or disagreement. This is happening presently in Taksim Square in Istanbul (Turkey), or in many cities across Europe facing an economic crisis. It also occurred in some Arab countries, during the "Arab Spring", when public squares were the stage where dissatisfaction with political decisions or systems was shown.

New critical approaches are necessary to the conception of public

位于雅典的罗马古市集是城市公共空间起源的典范之一
the Roman agora in Athens as an example of the orgin of urban public space

塔克西姆广场是伊斯坦布尔主要的城市公共空间，扮演着政治和公民象征的角色
Taksim Square, the main urban public space in Istanbul which performs a role as the political and civic symbol

下页中提到了三个项目并不仅仅是建筑，还是有用的城市设备，以及空间与公民之间的催化剂。它们的规划目的不仅仅是对其功能要求做出反应，还应该对周围的公共空间的活动举办起到决定作用，或者其本身便作为公共空间来使用。

交易大厅便是解决根特历史中心一个重要的城市失败案例问题的绝佳例子：这是一处政治决策将其转化为停车场的主要城市区域。根特市交易大厅的理念贯穿了16年，这一进程是十分独特的。它始于一次汽车停车场的设计竞赛，其上方带有一个公共广场，在这里，当地的Robbrecht en Daem建筑事务所与Marie-José Van Hee合作，选取了面向原始规划区对面的重要场地，且提出了空间使用的不同理念。他们的提议与建造停车场的要求截然相反，因此被迫取消。之后，公开的全民公投也迫使获奖的方案最终被放弃。

第二次比赛始于2005年，以Robbrecht en Daem建筑事务所与Marie-José Van Hee团队的提议作为参考。他们的项目被选取出来，这一大型的城市规划认可始于一座活动大厅的建设以及一个连续的中心广场规划的再认可。它包含一个综合项目，其中的城市连接作为一个统一体而存在：通勤者连接系统、历史中心游客环线以及当地居民的舒适性和可用性。

这座建筑在场地使用了双重建筑方法。在低层，通过一处绿化区，我们可以进入一座大咖啡馆、环形公园以及主要的基础设施。上层是一处干净的自由空间，市场大厅带有尖形的倾斜屋顶，这一形状来自于北欧意象，在这里，人们可以举办各种活动，并且进行互动。受到周围的天际线（不易模仿）的启发，这一抽象的当代木质外形采用了一种玻璃瓷砖来覆盖，创造了一次和历史遗产的有趣对话。

这是一个提前展示历史的嵌入结构，在这里，在公民对政治决策的态度方面，技术团队发挥了重要的作用，他们将广场返还给公民，以激活市中心的城市生活，促进公民关系。

棚式建筑展现了一个完全与众不同的环境，它是由Haworth Tompkins建筑师事务所设计的。这是一座用于容纳候补舞台的临时建筑，以对2014年竣工的伦敦国家剧院提供支援。这个项目是剧院最小的礼堂，即Cottesloe礼堂改建后的缩影。

这个项目是一座单体建筑，一个具有试验性的黑匣子，委员会成员、设计师以及剧院持有者都参与到它的设计理念当中。这一新场地就像是一个带有四座塔的巨大立方体，位于剧院广场中以及国家剧院的前方，以某种方式展现了其简化的几何形式。这个项目可以被认为是一个封闭体量，角落里带有四个烟囱，可用于自然通风，所采用的

space for the present, finding a balance between the extremes of adopting mediatic, trendy or typified models, or of applying radical activist proposals.

An urban public space should take into account facts such as globalization, mobility, increasing tourist flows or economic crisis situations, and deal with the diversified stakeholders–population, institutions, promoters, associations or individuals. Contemporary citizenship demands flexibility from the public space and a new civic role for the buildings. This can generate public participation and new civic actions and those spaces can be seen as new collective and relational spaces.

The three projects featured in the following pages are not mere buildings, but useful urban equipments and catalysts for spatial and civic relations. They are not just planned as a short response to a program, they are thought to generate action to the public space around or perform as the public space themselves.

The Market Hall project is a great example for solving an important urban failure in Ghent's historic center: a main urban area that indecision (or decision) of urban politics turned into a parking lot for years. The 16 years processus for the conception of the Market Hallin Ghent is peculiar. It started with a competition for a car parking with a public square on top, where local architects, Robbrecht en Daem Architecten together with Marie-José Van Hee adopted a critical position facing the original program, proposing different ideas for space usage. Their proposal countered the requested parking and was disqualified. After that, a public referendum obliged that the winning solution was also abandoned.

A second competition was launched in 2005, based on the proposal of the Robbrecht en Daem Architecten and Marie-José Van Hee team. Their project was chosen and a large urban requalification started with the construction of an events hall and the requalification of the contiguous central squares. It consists of a general project where city links are worked out as a whole: commuters connection systems, historic tourist circuit and comfort and usability for local residents.

The building has a double approach on the site. At a lower level, through a green area, we can access the grand café, the cycle park and the main infrastructure facilities. The upper level acts as a clean free space, with the market hall materialized in a sharp angled roof from the nordic imagery, where events and interactive activities can take place. Inspired by the surrounding skyline but far from being mimetic, this abstract contemporary wooden shape, covered with a sort of glass tiles, creates an interesting dialogue with the historical legacy.

This is an intervention that brings ahead history, where the technical team had an important role in their civic attitude facing political decision, allowing to return the square to the citizens, to activate city life in the center and to promote civic relations.

With a completely different context, The Shed, designed by Haworth Tompkins Architects is presented. It's temporary building that hosts a supplementary stage providing support to London's National Theater until April 2014, the time previewed for the renovation of the theater's smallest auditorium, the Cottesloe. The Shed is a singular building, an experimental black-box, that

1. Manuel Gauza, "Spaces: Collective or Relational (formerly Public) Space", *The Metapolis Dictionary of Advanced Architecture*, Barcelona: Actar, 2003
2. Zygmunt Bauman, *Globalization: The Human Consequences*, Cambridge: Polity, 1998, p.21

材料和设备都是完全可以再利用和再回收的。

位于Denys Lasdun的野兽派风格的国家剧院的灰色混凝土周边环境中，这座棚式建筑以其显眼的红色照亮了周围。不仅仅这种对比使其身份更加突出，其覆层材料，经过精细测量的粗糙木板也被应用在国家剧院的混凝土模架中。建筑的形状如同一个小型乐高积木堆成的城堡一样，嵌在国家大剧院之中。

我们立即被它吸引了。建筑的比例适宜，其简单性和自主性不允许任何部分与国家剧院的主楼相竞争，棚式建筑赢得了它自己的空间。这是一处智能的临时表演空间，在国家剧院的映衬下，是一处标志性的空间参照，成为一处场景。这座建筑还如同一座面临泰晤士河的军事塔，激发了人们的好奇心，召唤来了新访客，并且在主空间的入口前接待他们。

位于墨尔本的前音乐场地哈姆音乐厅由Roy Grounds在20世纪70年代设计，于1982年开放，在Grounds去世之后，其简单的几何外形，即带有野兽派/后现代主义特点的混凝土柱形与John Truscott设计的华丽室内形成了鲜明的对比。

最近，ARM建筑事务所与城市设计师和建筑师Peter Elloit一起接受委托，以重新审视艺术中心的总体规划———一座位于亚拉河南岸以及圣基尔达路沿线的综合性建筑（包括哈姆音乐厅），并且解决了音乐厅的主要问题。除了技术更新之外，室内空间进行了升级，空间所需的辅助设施也有所增设。所要解决的主要问题便是这样的一个事实：这座建筑作为城市界限而存在，与河岸没有视觉的连接，实物联系也非常少。

ARM建筑事务所在室外露台和与河流水位齐平的墩座墙之上设计了一个玻璃门厅，这个墩座墙是一个40m长的混凝土墙体，如同蛇一样蜿蜒，带有大型的玻璃洞口，来作为哈姆音乐厅河岸和餐厅一侧的第二入口。为了连接圣基尔达路（以及露台）和河岸散步道，建筑师建造了一个大型楼梯，且安装了一个24小时服务的公共电梯，电梯与较高体量处的墩座墙并置。这些新特点为场地和建筑带来了新的活力，将其自身转化为公共空间。

虽然这个新项目是一次拘泥于形式的尝试，但是其空间特征为当地的城市生活带来了全新的活力。它在视觉上变得更加具有渗透性，且允许游客欣赏来自或者望向墨尔本中央商务区的景色。在夜间，这些透明的效果形成了一个全新的光影游戏，使周围焕发活力。通过增加新的人行流线，这座建筑为游客和居民提供了一个全新的沿着亚拉河岸散步的机会。

included commissioners, designers and theater stakeholders in its conception. This new venue looks like a monolithic cube with four towers, placed in the Theatre Square, in front of the National Theatre, and somehow a simplified reflection of its geometry. It is conceived as a blind volume with four chimneys at the corners for natural ventilation, and constructed with materials and equipments that can be entirely re-used and recycled.

In its grey concrete surrounding – the Denys Lasdun's brutalist National Theater – The Shed shines with its highlighted red color. It is not just because of this contrast that marks its strong identity, but also its clad material, a rough timber carefully scaled like it was used in the National Theatre concrete formwork, and its shape, like a small Lego castle plugged to the National Theater.

We're immediately attracted to it. The proportion is harmonious and its simplicity and autonomy don't allow any competition with the main National Theater building, The Shed gains its own space. It is an intelligent ephemeral space for performances that performs itself in the local, as an iconic spatial reference with the National Theatre as scenario. Like a military tower facing the Thames River, it stimulates curiosity, calling new visitors and receiving them before the entrance in the main space.

Hamer Hall in the premier concert venue in Melbourne, was designed by Roy Grounds during the 1970's and opened in 1982, after Grounds' death. Its simple geometry, a concrete cylinder shape with a brutalist/late modernist character, contrasted with the exuberant interiors designed by John Truscott.

More recently, ARM Architecture, together with urban designer and architect Peter Elliot, was commissioned to review the masterplan of the Arts Center – a complex of buildings in the Yarra River's southbank and along the St Kilda Road that includes the Hamer Hall – and solve major problems of the concert hall. Besides the technical update, improvements of interior spaces and support facilities that the space needed were also made. The main problem to work out was the fact that the building acted as an urban barrier, with no visual connection and a poor physical relation with the riverside.

ARM Architecture designed new glazed foyers over an exterior terrace and a new podium on the river level. This podium is a 40 meters concrete wall, waving as a serpentine, with large glass openings for a second entrance to the Hamer Hall from the riverside and for restaurants. To connect St Kilda Road–and the terrace–with the riverbank promenade, a large stair was projected and a 24 hours lift for public use juxtaposed to the new podium in a higher volume, was installed. These new features bring a new life to the site and to the building, which turns itself in public space. Although this new project is a formalist exercise, its spatial characteristics trigger a new dynamic to local urban life. Being more permeable visually, it allows sightseeing from and to Melbourne's CBD. At night, these transparencies generate a new lightening play and animate the surroundings. By incorporating new pedestrian circulations, it offers tourists and citizens a new walk opportunity along Yarra's riverfront. Paula Melâneo

城市改造 公民参与 | Urban How Civic Engagement

交易大厅

Robbrecht en Daem Architecten + Marie-José Van Hee

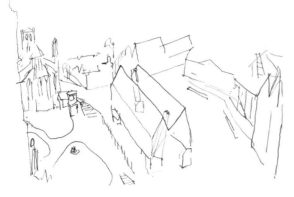

1 圣尼古拉斯教堂	5 科伦市场广场	9 公园
2 钟楼	6 Saint Baafsplein广场	10 交易大厅
3 圣贝茨大教堂	7 Gouden Leeuwplein广场	11 雕塑
4 市政厅	8 Poeljemarkt广场	
1. Saint Nicholas Church	5. Korenmarkt Square	9. park
2. belfry	6. Saint Baafsplein Square	10. Market Hall
3. Saint Baats Cathedral	7. Gouden Leeuwplein Square	11. sculpture
4. city hall	8. Poeljemarkt Square	

在1913年世界博览会经历的两次拆除运动以及20世纪60年代永远没有建成的市政中心之后，根特市的历史中心经过几十年的洗礼，逐渐褪去繁华，成为一座荒凉的停车场，位于毗邻的三幢哥特式的尖顶塔楼之间。在1996年与2005年间连续举办的两次建筑竞赛中，Robbrecht en Daem建筑事务所和Marie-José Van Hee事务所提出了其规划，并且颠覆了最初的竞赛规则。这个项目规划没有提供一处开放的、用于举办活动的空间，而是在此处精心设置了一座交易大厅，来修复这一场地缺陷，恢复旧城区域，且被公众所接受。

这座建筑位于Gouden Leeuwplein的Poeljemarkt广场与一条低矮的、新建的绿化带之间，这条绿化带连接着大厅下方的"小酒馆"、自行车公园和公共卫生间。建筑用地面积为24 000m²，位置较为显著，设置合理。与圣尼古拉斯教堂、钟楼、大教堂相比，这座建筑和较低建筑群的高度一致，如临近的市政厅。

建筑的内部较为城市化，室内采用双重调音木质天花板，将游客拥抱其中，天花板上的小窗户射进点点阳光。事实上，整座建筑的室外都似乎通过使用木材（其饰面朴实无华）来向高贵的历史石建筑致敬。玻璃外围护结构对木材起到了保护的作用，使照射在其中的光线经过天空的反射较为柔和，且与天空融为一体。

大型缓冲地带吸纳雨水，小酒馆所采用的低能耗原则，绝对自然的材料、公共交通工具的推广，以及完美的视觉（利用之前的空间结构来为历史中心赋予新价值）都是把将来的可持续性具体化的一部分。这座根特的中心——交易大厅将再次成为市民社交场所。

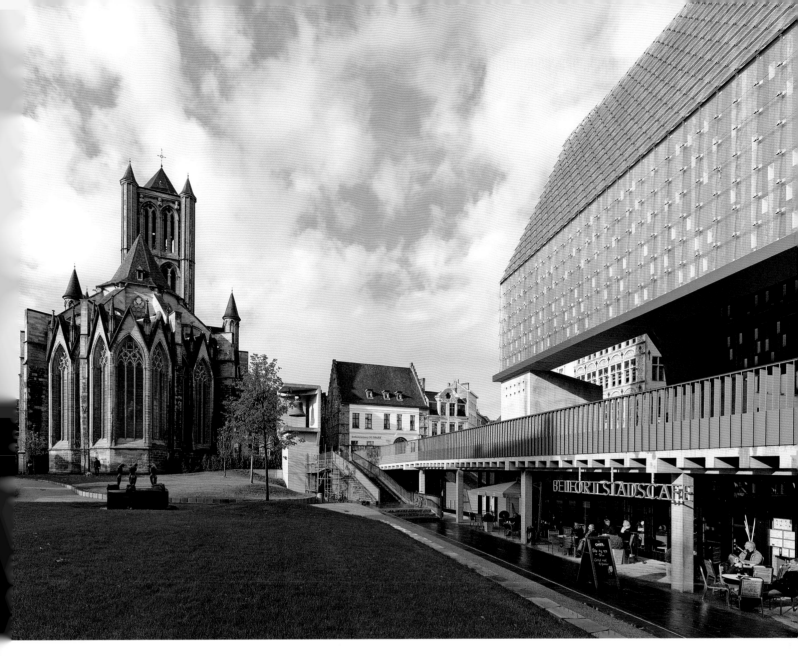

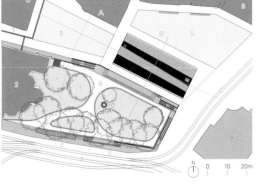

1 Klein Turkije街	1. Klein Turkije Street
2 圣尼古拉斯教堂	2. Saint Nicholas Church
3 Gouden Leeuwplein 广场	3. Gouden Leeuwplein Square
4 交易大厅	4. Market Hall
5 Poeljemarkt广场	5. Poeljemarkt Square
6 公园	6. park
7 Catalonie街	7. Catalonie Street
8 市政厅	8. city hall
9 钟楼	9. belfry
10 雕塑	10. sculpture

C-C' 剖面图 section C-C'

Market Hall

Following two demolition campaigns for a world exhibition in 1913 and an administrative center never built in the 1960s, Ghent's historic heart degenerated for decades into a desolate parking lot between a suite of three adjoining Gothic towers. In two consecutive competitions between 1996 and 2005, Robbrecht en Daem Architecten and Marie-José Van Hee proposed their own program, countering the initial competition requirement. Rather than just providing an open space for events, they sought, by meticulously positioning a market hall, to rectify this deficiency and reinstate the presence of old urban areas that had become unrecognizable.

The building positions itself between Poeljemarkt, Gouden Leeuwplein, and a new lower "green" connecting to the "brasserie", bicycle park and public toilets below the hall. Although the building clearly occupies a position on the 24,000m² site, it fits in well. Compared to St. Nicholas Church, Belfry and Cathedral, it assumes the heights of a lower group of buildings such as the adjacent town hall.

As an urban interior, the inside embraces the passer-by with a dual modulated wooden ceiling, whose small windows scatter light inwards. The exterior, of the entire building in fact, seems to assume a respectful role relative to the nobler historic stone buildings, by using a wooden, almost humble, finish. A glass envelope protects the wood and provides a soft shine, with the sky reflected, integrated.

Large buffer basins to absorb rainwater, principles of low energy consumption for the brasserie, use of truly natural materials, the contribution of public transport and a clear vision about giving new value to the historic center with its old spatial structures, are just parts that broadly flesh out "sustainability" for the future. The center of Ghent will again become a social spot for people.

项目名称：Market Hall
地点：Korenmarkt and E. Braunplein, City of Ghent
建筑师：Robbrecht en Daem Architecten, Marie-José Van Hee
项目建筑师：Bert Callens, Wim Menten
合伙人：Jan Baes, Tom Broes, Katrien Cammers, Axel Clissen, Mattias Deboutte, Petra Decouttere, Arne Deruyter, Linde Everaerd, Trice Hofkens, Gert Jansseune, Daniël Libens, Carmen Osten, Filip Reumers, Sofie Reynaert, Miriam Rohde, Johannes Robbrecht, Marilù Sicoli, Gert Swolfs, Pieter Vanderhoydonck, Kathy Vermeeren, Caroline Voet, Wim Walschap
结构工程师：BAS, Dirk Jaspaert
服务工程师：Boydens
技术工程师：Studiebureau Boydens
基建工程师：Technum-Tractebel Engineering
景观建筑师：Wirtz International
甲方：City of Ghent, VVM De Lijn & TMVW
功能：events hall, public functions and construction of streets and squares
用地面积：24,000m²
总建筑面积：625m²
有效楼层面积(低层)：1,400m²
设计时间：1996
竣工时间：2012
摄影师：©Marc De Blieck (courtesy of the architect)

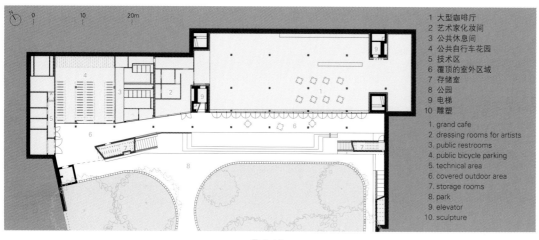

一层 first floor

1	大型咖啡厅
2	艺术家化妆间
3	公共休息间
4	公共自行车花园
5	技术区
6	覆顶的室外区域
7	存储室
8	公园
9	电梯
10	雕塑

1. grand cafe
2. dressing rooms for artists
3. public restrooms
4. public bicycle parking
5. technical area
6. covered outdoor area
7. storage rooms
8. park
9. elevator
10. sculpture

1. ceiling of European oak
2. steel structure
3. plywood deck, waterproofing
4. facade, hardwood
5. facade cladding patterned toughened glass
6. roof light

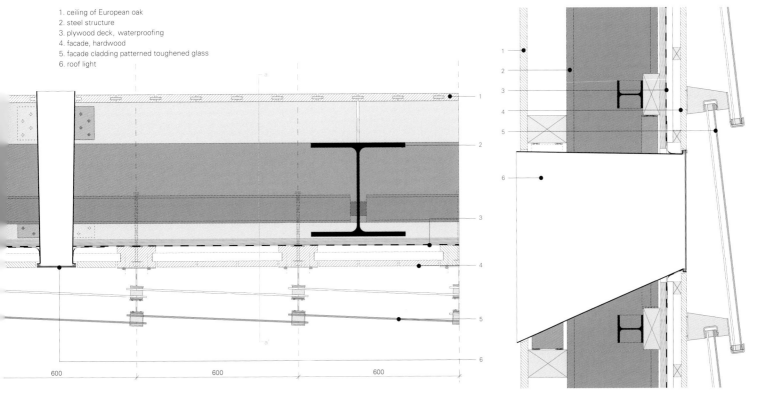

屋顶平面详图 roof plan detail

a-a' 剖面图 section a-a'

棚式建筑
Haworth Tompkins Architects

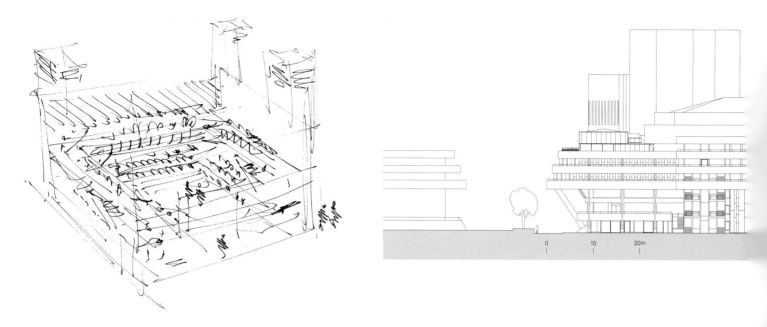

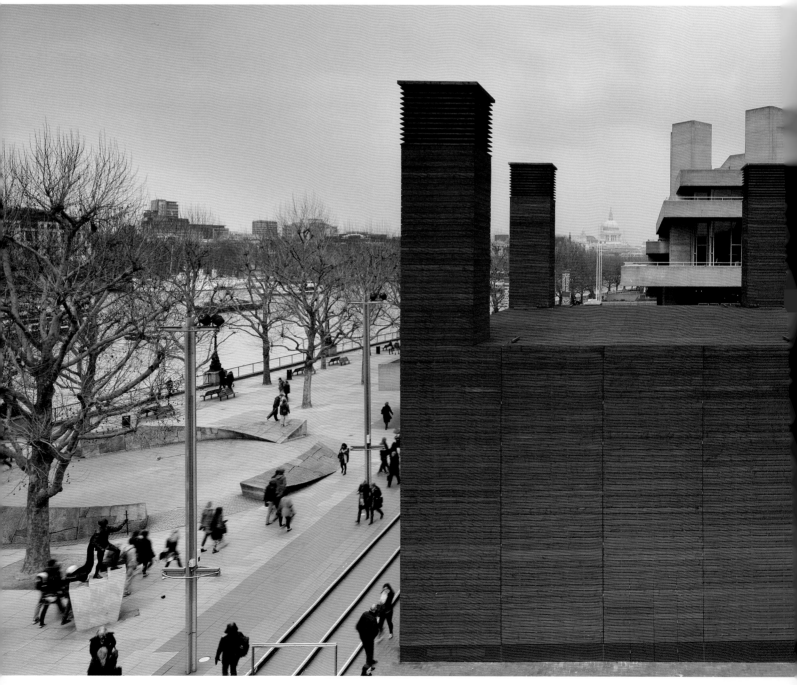

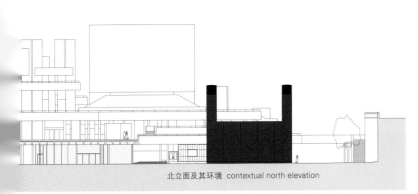

北立面及其环境 contextual north elevation

这座建筑是一处为伦敦南岸的国家大剧院所设置的临时演出地点，在国家大剧院的重建（也是由Haworth Tompkins建筑事务所设计）期间，科特索海滩被关闭，而这座建筑将会容纳国家大剧院三分之一的礼堂。

由国家大剧院主管Nicholas Hytner指定的棚式建筑的艺术功能，促进了艺术创作的范围，给予了探索新的国家大剧院建造方式的机会。同样，这座棚式建筑也成为建筑设计团队的试验地。

按照Haworth Tompkins建筑事务所和剧院顾问Charcoalblue的规划，这座建筑的设计和建造仅仅用了一年多一点的时间。共同工作在一起的建筑设计师、国家大剧院、剧院建造者之间的合作成果，犹如剧院的一场演出，而非传统意义上的建筑项目。

由于其临时场馆的属性，如同Haworth Tompkins建筑事务所早期设计的其他临时场馆——盖恩斯伯勒工作室的阿尔梅达剧院与国王十字车站，这座建筑更多的是被看成一个活动，或是一个装置，而非一座建筑。它是伦敦南岸的一个充满活力的嵌入结构，在12个月的服务期内让观众为之倾倒。

这座建筑占领了剧院的广场空间，在国家大剧院的正面，毗邻泰晤士河。其形式较为简单，设有一个拥有225个座位的礼堂，材料为原钢和胶合板。粗锯的木材覆层适和应用在国家大剧院具有标志性的模板纹表面的混凝土中，而礼堂的立体感与其角落处的塔楼则与国家大剧院鲜明的几何外形相得益彰。建筑的临时门厅与国家大剧院室外露台下方的空间隔离开来，并且与现在的门厅直接相连。耀眼的红色除了门和窗户外，都对整座建筑进行了覆盖，大胆地宣告了其到来，与国家大剧院的几何混凝土形式形成对比，呈现出令人震撼的神秘景象。

这座建筑也体现了Haworth Tompkins建筑事务所在建造的项目中，采用可持续性的方法来建造剧院所迈出的一步。建筑的材料都是100%可回收的，其配套的坐椅也是可循环利用的。该建筑能够进行自然通风，其四个塔楼通过建筑的不同形式来将风引入建筑内。

剧院主管Steve Tompkins表示："这次合作为双方共同探索临时公共性结构的建造方法提供了绝佳的机会，改变了我们以往对地点与布局的概念。我们希望这座建筑是一座娱乐性的建筑，但是更希望它能引人思考，且对伦敦南岸的永久性文化建筑提出调整，或者进行互补。

Charcoalblue剧院的执行合伙人Andy Hayles说道："Haworth Tompkins建筑事务所是一个了不起的设计团队，他们理解观众的需求，鼓励团队创新，努力建造一座人们乐用、观众愿来的建筑。非常荣幸能与Steve及其团队合作来建造这座国家大剧院的临时项目，我们期盼着项目投入使用时来自剧院的反馈意见，希望未来能够再度合作。"

The Shed

The Shed is a temporary venue for the National Theater on London's South Bank. It will give the National Theater a third auditorium while the Cottesloe is closed for a year during the National Theater future redevelopment, also designed by Haworth Tompkins.

The artistic program for The Shed, recently announced by the director of the National Theater, Nicholas Hytner, pushes creative boundaries, giving the National Theater the opportunity to explore new ways of making theater. In the same way, The Shed has been a test bed for experiment by the architectural design team. Conceived by Haworth Tompkins and theater consultants Charcoalblue, it was then designed and built in little more than a year, a collaborative process between the building designers, the National Theater, and theater-makers who will work in the space, in a way that more closely resembles a theater show than a conventional construction project.

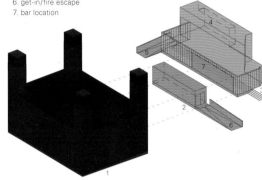

1 礼堂
2 发挥连接作用的体量
3 门厅
4 双高的入口区域
5 通往主入口的斜坡
6 入口区/安全出口
7 酒吧

1. auditorium
2. linking volume
3. foyer
4. double height entrance area
5. ramp to main entrance
6. get-in/fire escape
7. bar location

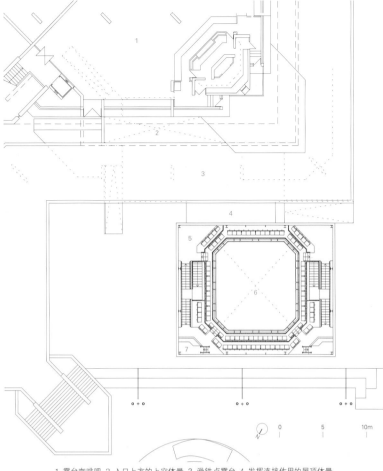

1 露台咖啡吧 2 入口上方的上空体量 3 滑铁卢露台 4 发挥连接作用的屋顶体量
5 位于上层的礼堂 6 舞台上方的上空体量 7 灯光和音效控制室
1. terrace cafe 2. void over entrance 3. Waterloo terrace 4. linking volume roof
5. auditorium on upper level 6. void over stage 7. light and sound control booths

二层 second floor

1 主入口 2 现存的利特尔顿式门厅 3 门厅 4 酒吧 5 更衣室 6 礼堂的大堂
7 礼堂 8 现存的咖啡吧 9 剧院大道 10 剧院广场 11 女王通道 12 上空体量
1. main entrance 2. existing Lyttelton foyer 3. foyer 4. bar 5. dressing room 6. auditorium lobby
7. auditorium 8. existing cafe 9. theatre avenue 10. Theatre Square 11. Queen's Walk 12. void over

一层 first floor

项目名称：National Theatre "The Shed"
地点：South Bank, London, SE1 9PX
建筑师：Steve Tompkins, Paddy Dillon, Shane McCamley
结构工程师：Flint & Neill Ltd
服务工程师：Ingleton Wood LLP
承包商：Rise Contracts Ltd
甲方：National Theatre
用途：auditorium
室内有效楼层面积：628m²
造价：GBP 1.2m
施工时间：2012.9—2013.2
摄影师：
©Helene Binet (courtesy of the architect) - p.49, p.51
©Philip Vile (courtesy of the architect) - p.44~45, p.46~47, p.50, p.52, p.53

Its temporary nature, building on Haworth Tompkins' earlier temporary projects like the Almeida Theater at Gainsborough Studios and King's Cross, permits a structure that can be seen less as a building than as an event or installation – a vibrant intervention on London's South Bank designed to entrance, and sometimes bewilder, passers – for a period of twelve months.

The Shed occupies the Theater Square, at the front of the National Theater, beside the river. Its simple form houses a 225-seat auditorium made of raw steel and plywood. The rough-sawn timber cladding refers to the National Theater's iconic board-marked concrete, and the modeling of the auditorium and its corner towers complement the bold geometries of the National Theater itself. A temporary foyer has been carved out from the space beneath the National Theater's external terraces and provides easy connection to the existing foyers. The Shed's brilliant red color covering the entire mass of a form without doors or windows, announces its arrival boldly against the geometric concrete forms of the National Theater, giving it a startling and enigmatic presence.

The Shed also represents another step in Haworth Tompkins' ongoing project to research sustainable ways of making theaters. Built of materials can be 100% recycled and fitted out with re-used seating, The Shed is naturally ventilated, with the four towers that draw air through the building providing its distinctive form.

Director Steve Tompkins said: *"This collaboration has been a wonderful opportunity to explore the ways in which temporary public buildings can alter our perceptions of places and organizations. We hope The Shed will be seen as a playful but thoughtful building, both challenging and complementary to the permanent cultural architecture of the South Bank."*

Andy Hayles, Managing Partner of Charcoalblue said: *"Haworth Tompkins are quite extraordinary designers. They understand the needs of theater users, encourage and elicit team-wide innovation, and manage to create buildings that theatre people love to use and audiences love to visit. We are honored to have jointly conceived The NT Shed with Steve and his team, and look forward to sharing the feedback from the National as they use the building to further inform our future collaborations."*

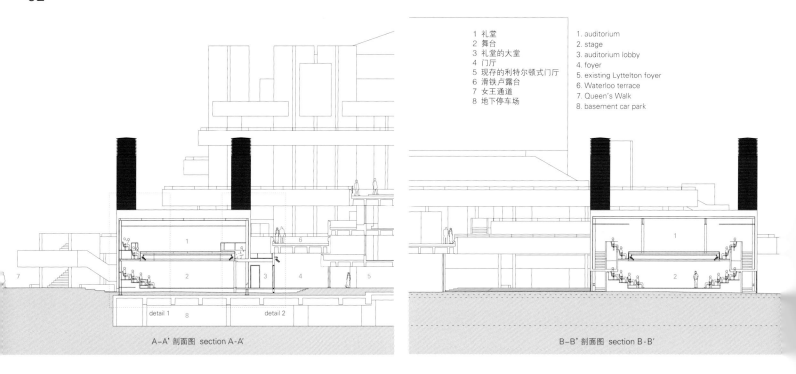

1 礼堂	1. auditorium
2 舞台	2. stage
3 礼堂的大堂	3. auditorium lobby
4 门厅	4. foyer
5 现存的利特尔顿式门厅	5. existing Lyttelton foyer
6 滑铁卢露台	6. Waterloo terrace
7 女王通道	7. Queen's Walk
8 地下停车场	8. basement car park

A-A' 剖面图 section A-A'

B-B' 剖面图 section B-B'

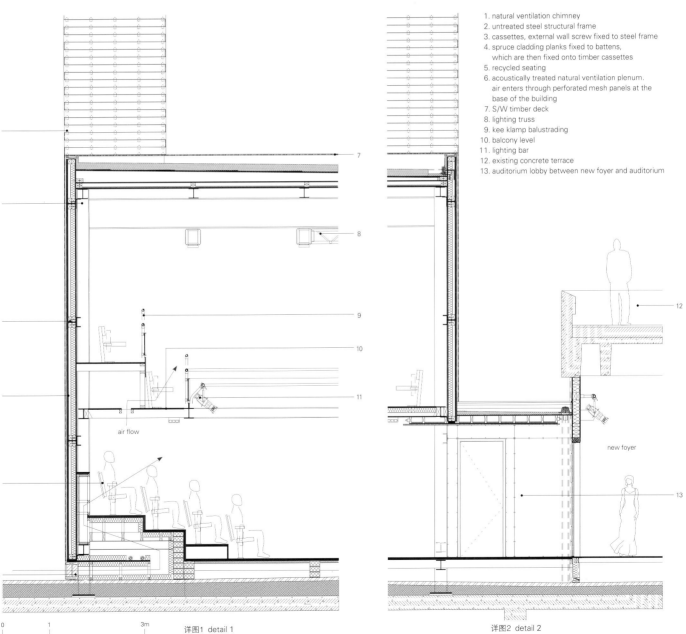

1. natural ventilation chimney
2. untreated steel structural frame
3. cassettes, external wall screw fixed to steel frame
4. spruce cladding planks fixed to battens, which are then fixed onto timber cassettes
5. recycled seating
6. acoustically treated natural ventilation plenum. air enters through perforated mesh panels at the base of the building
7. S/W timber deck
8. lighting truss
9. kee klamp balustrading
10. balcony level
11. lighting bar
12. existing concrete terrace
13. auditorium lobby between new foyer and auditorium

详图1 detail 1   详图2 detail 2

城市改造 公民参与 | Urban How Civic Engagement

墨尔本艺术中心的哈姆音乐厅
ARM Architecture

哈姆音乐厅的潜在设计理念来自罗伊·格朗兹和约翰·特斯科特两人相互冲突的主题的力量。建筑师的设计任务包含对艺术中心管理区整体规划的检阅，以及其与宏伟的南岸的关系。

建筑师的任务是在逐步协调与各个甲方参与者之间的矛盾和协议后，来开放哈姆音乐厅，将其与河流融为一体，成为21世纪游客的聚集地。而建筑师的工作便是修正其地位，使其成为一座主要的、服务于市民的建筑。这座音乐厅代表了墨尔本的两种艺术篇章：城堡的建筑历史与矿藏的建筑历史；宫殿的戏剧与珍宝洞的戏剧。建筑师采用这两种叙述，将之重新解读，并且在哈姆音乐厅中加入自己的故事，新建的河畔部分的设计灵感就来自新兴的几何学。

本次重建得到墨尔本维多利亚遗产委员会的批准。在整个重建过程中，建筑师极力保护音乐厅许多独特的遗址，尽量保留了罗伊·格朗兹的基础建筑，并且尽可能地保留约翰·特斯科特的内部原有结构。

重建工作修复了哈姆音乐厅的音响设施，使其音响效果堪称世界顶尖水平，采用先进的技术来安装艺术舞台和后台设施。重新的设计使观众的舒适性得以加强，并且增设了门厅、卫生间以及室外进行改造的休息室。这一改造步骤创造一个更加向外开放的场地，使其更加方便人们进入，更具有吸引力。

重建保留了现有的鼓形结构，但是其管理区的临海一面做了较大的改动。新外形产生于格朗兹的灵感，即罗马圣安吉洛城堡。整座建筑采用装饰性混凝土建成，使人们想起罗马废墟的具有乡土风格的基础。场地的布局强化了建筑周围现存的轨道痕迹。

建筑师的任务重新斟酌FJMT早些时候完成的总体规划。项目资金主要用于重塑高水准的哈姆音乐厅形象、攻克改建音乐厅的技术难题以及出入不便的问题。很快，团队就开始面临要解决改建中的主要功能性挑战，包括提高音乐厅与周围环境的融合度，尤其是与圣基尔达街和河流之间的融合度，修复音乐厅大部分音响设备，大量地改进操作技术，新建餐饮部以增加三分之一的收入以及提高舒适性等。这一系列改建在整个商业案例中都是隐形存在的，从未实实在在地算入项目预算中，他们却完成了。

ARM建筑事务所与设计团队协调合作，这个设计团队包括Peter Elliott（城市设计）、TCL（景观）、Kirkegaard Associatess和Marshall Day（声学）、Aurecon（服务&结构）以及Shuler Shook（剧院设计师）。

该项目通过签订联盟合同来交付，也就是说，建筑师与项目交付方，如业主、施工人员以及设计师地位同等。哈姆音乐厅的改建费用由维多利亚州预算支付，但合同规定的施工时间期限太短，不切实际。出乎意料的是，建筑的规模和功能完美地实现了，远远超出常规的"商业范畴"和甲方预期的效果。

项目名称：Hamer Hall, Arts Center Melbourne
地点：26-28 Southgate Ave, Southbank, VIC 3006 Australia
建筑师：ARM Architecture
项目团队：Ian McDougall, Stephen Ashton, Howard Raggatt, Neil Masterton, Peter Bickle, Stephen Davies, Jonothan Cowle, Andrea Wilson, Rhonda Mitchell, Doug Dickson, William Pritchard, Paul Buckley, Justin Fagnani, Sarah Lake, Tom Denham, Matthew Ginnever, Allira Davies, Tim Brooks, Asako Miura, Andrew Lilleyman, Aaron Poupard, Andrew Ta, Deborah Rowe, Jason Lee, Ken Billan, Lee Lambrou, Mark Raggatt, Martine, De Flander, Monique Brady, Mordechai Toor, Natalie Lysenko, Sarah Box, Sarah Lake, Simon Shiel, Tobi Pederson, Tom Denham, Tom Marsh
景观建筑师：Taylor Cullity Lethlean
城市设计：Peter Elliott Architecture
音响顾问：Kirkegaard Associates & Marshall Day
剧场顾问：Schuler Shook 甲方：Major Projects Victoria
遗产顾问：Bryce Raworth & Phillip Goad
结构与服务：Aurecon 人行道建模：Arup
灯光顾问：Lighting Design Partnership
质量监理：Rider Levett Bucknell
用途：public architecture – theater
有效楼层面积：16,000m² 造价：USD 128.5m
竣工时间：2012.7
摄影师：©John Gollings (courtesy of the architect)

Hamer Hall of the Arts Center Melbourne

The underlying ideas behind the design come from the strength of the conflicting thematics by Roy Grounds and John Truscott. The architects' commission comprised a review of the master plan for the whole of the Arts Centre precinct and its relationship to the greater Southbank.

The architects' mission, gradually refined through conflicts and agreements with the various client participants, was to open up Hamer Hall, integrating it with the river, and make it a 21st Century venue. Underlying their job was to repair its standing as a major civic building. The building represents two artistic discourses in Melbourne: the architectural story of the castle and the mine: the theatrical story of the palace and the cave of jewels. The architects took these two narratives, reinterpreted them and added their own voice to the story of Hamer Hall. The new riverside section is inspired by a new geometry.

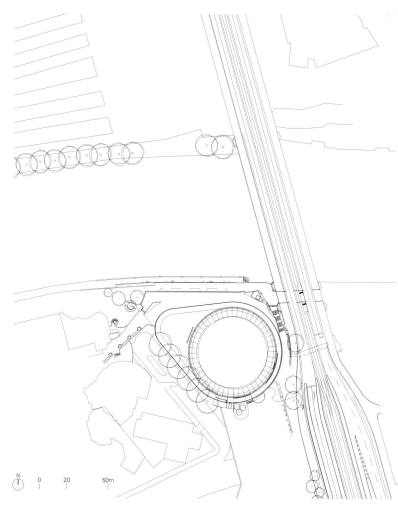

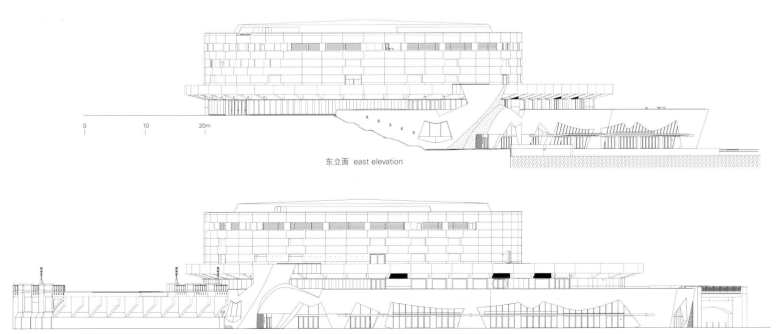

东立面 east elevation

北立面 north elevation

The redevelopment was approved by Heritage Victoria. Throughout the redevelopment, they took great care to preserve many unique heritage aspects of the Hall, retaining Roy Grounds' base architecture and as much as possible of John Truscott's original interiors.

The redevelopment repairs HH's performance acoustics, to world standards, installs state of the art stage and back of house facilities. The patron amenity is completely redesigned, with increased foyer area, toilet numbers and lounges with exterior modifications that create a more outward facing venue to make it more accessible and inviting to the public.

The existing built form of the drum has been retained but radical changes were made to the riverside section of the precinct. The new form draws its influences from Grounds' muse, the Castel San Angelo, Rome. Built in off-form concrete it recalls the rusticated base of the Roman ruin. The layout reinforces the existing tracks around the building.

The architects' commission was to rethink the master plan completed earlier by FJMT. The funding for the project had been primarily focused on high level ambitions for re-branding Hamer Hall and repairing theater technology and poor access. Quickly the team had honed in on the main programmatic challenges to be addressed. These included improved integration with the Hall's surroundings, especially St Kilda Rd and River side, substantial Hall acoustic repair, substantial operating and technical improvements, a third revenue stream through new F&B spaces and improved patron amenity. While this was latent in the overall business case, it had never been physically translated into a budget. They did that.

ARM coordinated the design team which included Peter Elliott (Urban Design), TCL (Landscape), Kirkegaard Associatess and Marshall Day (acousticians), Aurecon (Services & Structure) and Shuler Shook (theatre designers).

The project was delivered through an Alliance contract, which means that the architects were an equal member of the project delivery team comprising owner, constructor and designer. Hamer Hall reborn was delivered within the State's target budget, and within the prescribed but unrealistically short time frame. Anecdotally, the scope and functionality have been achieved well beyond "business as usual" and the expectations of the client.

ARM Architecture

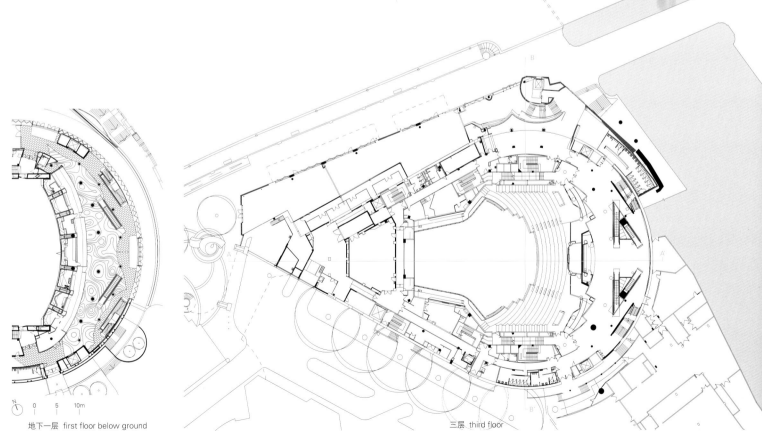

地下一层 first floor below ground

三层 third floor

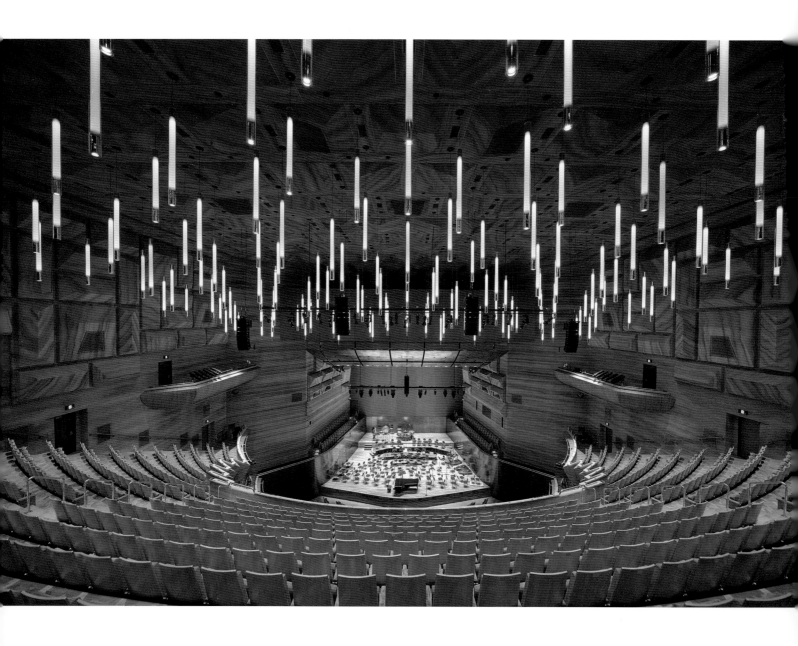

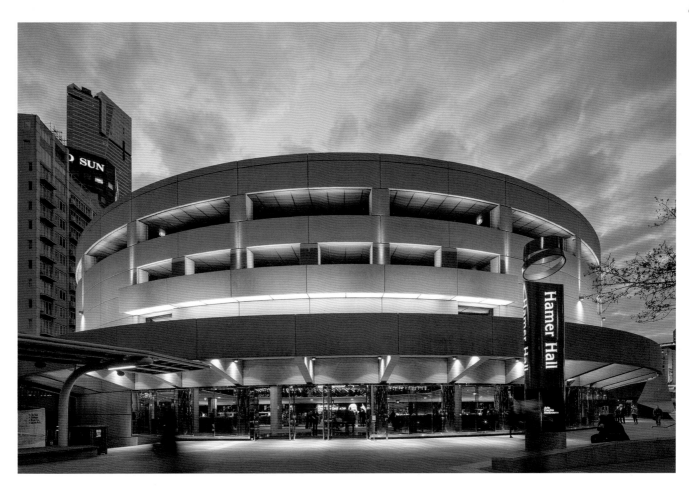

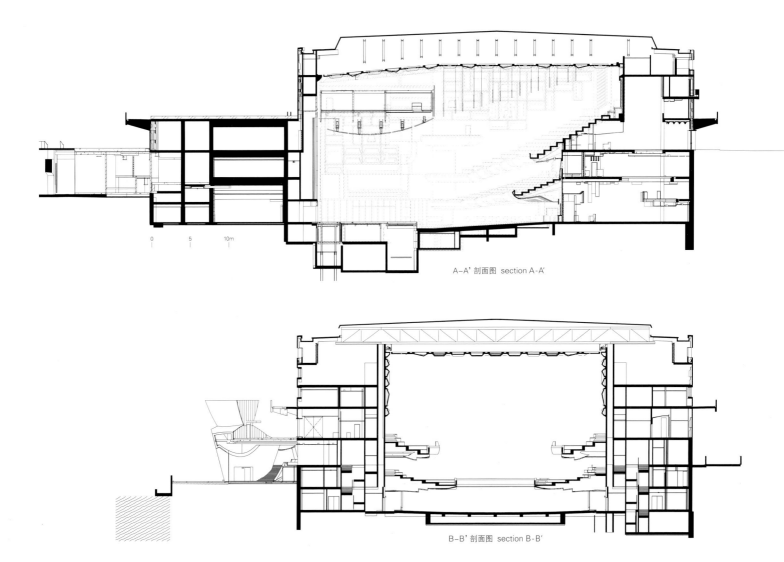

A-A' 剖面图 section A-A'

B-B' 剖面图 section B-B'

微工作·微空间
Minor Works

大多数人都很清楚自己希望从工作场所中获得怎样的工作效果。当然，工作的类型不同，效果也千差万别。但是每个人都可能希望自己的工作场所能有充分的采光、良好的隔音效果，且相对于外界来说具有一定的私密或开放性。下列的八个工作场所特点各异，有的身藏居民区，有的位于公共建筑内，建筑师们用各种方式来处理私密和开放的难题，同时也提出了一些问题：为了高效工作，办公室和建筑的其他功能区应分隔至何种程度？办公室应该离外界多远？将办公室与外界隔离的最好方法是什么？下列的最近完工的八个小型工作空间在这些问题上，都有各自的解决方案。

Most people have a clear idea of what they want from a workspace. Of course this will vary according to the type of work, but their wish list is likely to include appropriate lighting and acoustics and a certain amount of either privacy or openness to the outside world. The eight workspaces featured here are mostly surrounded by other functions. They range from a study space in a residential studio, to the offices of public buildings and their architects have approached the privacy and openness issue in a number of ways. They pose a number of questions: to what extent does the office need to be separated from the other functions of the building for work to take place there effectively? How far should it be cordoned off from the outside world? And what is the best way to do that? These eight recently completed small workspaces have their own perspective on these questions.

Stardom娱乐公司办公室的重建/D·Lim Architects
卡萨雷克斯办公室/FGMF Arquitectos
森林中的办公室/SUGAWARADAISUKE
Rubido Romero基金会/Abalo Alonso Arquitectos
马德里博坦基金会的新办公室/MVN Arquitectos
Kirchplatz办公室和住宅/Oppenheim Architecture+Design
Torus/N Maeda Atelier
莫托萨布公寓sYms/Kiyonobu Nakagame Architect and Associates

微工作·微空间/Alison Killing

Stardom Entertainment Office Remodeling/D·Lim Architects
Casa Rex Office/FGMF Arquitectos
Office in Forest/SUGAWARADAISUKE
Rubido Romero Foundation/Abalo Alonso Arquitectos
The New Offices of the Botín Foundation in Madrid/MVN Arquitectos
Kirchplatz Office and Residence/Oppenheim Architecture+Design
Torus/N Maeda Atelier
Motoazabu Apartment sYms/Kiyonobu Nakagame Architect and Associates

Minor Works/Alison Killing

微工作·微空间

工作间可以设计得非常细致，员工对工作间的要求不可避免地要随手头的任务的不同而变化，当然，有时也随心情而变化。这种情况下，设计师就必须密切关注采光和隔音效果，以确保高效地工作，同时要保证必要的私密和开放空间，为员工们提供舒适的工作环境。小型工作间必然受到周围功能区的干扰，因此，隐私和隔离是首要考虑的问题。为了高效工作，办公区域和建筑的其他功能区应分隔至何种程度？办公室应该离外界有多远？将办公室与外界隔离的最好方法是什么？

Kiyonobu Nakagame联合建筑师事务所设计的莫托萨布公寓sYms解决了如何在一栋房屋内分隔格子间以达到最佳格局的难题。

用作工作间的公寓面积较小，拥有足够的开放性的空间是弥足珍贵的。对于一处为不同功能而服务的专用空间来说，其标准的方案是建造隔墙，以划分空间，但是，这种做法使房间成为一个常规的公寓。

在本案中，设计师们尽可能地使空间开阔。这栋建筑夹在两栋相邻的建筑中间，其前墙和后墙几乎全部采用玻璃来覆盖，从街上看向屋内或者从屋内向外看街道的视线都很好。工作区、睡眠区、餐饮区、休闲区通过地面和天花板上成对角线形的台阶分隔开，天花板上的阶梯与地面的阶梯相互垂直，使这四处区域将整个大房间分割成具有理想高度的功能区。

这是一个明智的方法，把不同的功能整合在一个狭小的楼层空间

Minor Works

Workspaces can be delicate things. Their occupants' requirements will inevitably vary according to the task at hand and also perhaps, their mood. For the designer this tends to mean paying careful attention to lighting and acoustics to allow the work to be done effectively, but also providing the privacy or openness necessary for people to work comfortably. In a small workspace, surrounded by other functions, the questions of privacy and separation come to the fore: to what extent does the office need to be separated from the other functions of the building for work to take place there effectively? How far should it be cordoned off from the outside world? And what is the best way to do that?

Motoazabu Apartment sYms by Kiyonobu Nakagame Architects and Associates has distilled the question of how to create a separate working space within a house down to its purest form. With the studio apartment occupying only a relatively small area, space and a feeling of openness to go with it, were at a premium. The standard solution to the problem of how to provide a dedicated space for different function would be building walls to define rooms, but that would have created a poky apartment.

Instead, the architect has left the space as open as possible. The building is sandwiched between two neighbors, but the front and back walls are almost entirely glazed, with views out, as well as in, from the street. The different spaces of the house, for working, sleeping, eating and relaxing are demarcated by diagonal steps on the floor and ceiling. The ceiling step runs perpendicular to the one in the floor, so that together they divide the larger room into ideal height to function as a seat.

It is an intelligent approach to combine numerous functions within a tiny floor place, but it's possible that this only really works when the apartment has a single occupant. There is no privacy between the different spaces, whether visual or acoustic, which would make the study, in particular, less than ideal in an apartment shared by more people. It takes a peculiar attitude to privacy, where even the walls to the bathroom and WC are made from transparent glass.

The Torus by N Maeda Atelier also combines the work and residential function, with the owner and their family living above their grooming salon for dogs. The two functions are strictly separated, with only the entrance of the house on the first level and the rest of the dwelling packed into an opaque white volume resting on top of the glass clad space of the shop. It is an incredibly introspective house, above the outward facing, public salon.

For the obvious reason of attracting customers, the shops needs to open up to the outside. The glass and translucent mesh which makes up the walls on the first floor, appears at first to be free form, but the "bites" which have been taken out are actually to accommodate entrances, parking, or space for equipment. The space at the back of the building is for the dogs to run around.

The living space on the second and third floor is completely removed from the working world at street level. Climbing the spiral staircase which connects the two, one enters a different world, leaving work behind. The "Torus"'s nickname comes from its donut-like form. The building has a central void, capped by a large roof light which provides much of the daylight available to the

照片提供：©Oppenheim Architecture + Design (Borje Müller)

Kirchplatz办公室和住宅中较为现代化且统一的办公空间，由一座18世纪的农舍改造而成

modernized and united office space of Kirchplatz Office and Residence, converted from 18 century farmhouse

内，但是，只有在单独的业主方拥有公寓的情况下，这种办法才奏效。无论耳听还是眼观，不同的区域间都没有隐私，尤其是多人在一间公寓内学习时，这并不是一个理想的场所。人们处在这样的地方对隐私问题要有特殊的心态，因为盥洗室和卫生间的墙都是透明玻璃。

N Maeda Atelier建筑事务所设计的Torus建筑同样将工作和居住的功能完美地组合起来，业主和家人居住在楼上，楼下是宠物狗美容店，工作和居住区严格区分，房屋一层只有一个出入口，其他部分均设置在一个不透明的白色体量之内，依附在商店的玻璃覆层上。这栋建筑位于面向外部的公共美容店之上，向内开放，给人以难以置信的感觉。

为了吸引顾客，店面必须面街，一楼的玻璃和半透明网格墙第一眼看上去形式较为随意，但是被移走的"网眼"实际上可以容纳出口、停车场或设备间，楼后的空地则供狗儿们撒欢。

二层和三层的居住区与和街道水平高度保持一致的工作区完全分开。沿着连接这两个楼层的螺旋形楼梯拾阶而上，人们进入了一个完全不同的世界，将工作抛在脑后。Torus建筑的这一绰号来自于房屋圆环的形状，房屋中央为上空体量，巨大的屋顶采光极好，阳光直接照进建筑的上层房间。

由于不同组的人使用，Oppenheim建筑和设计事务所在Kirchplatz设计的办公和居住混合的建筑中将办公和居住两种功能清晰地划分出来，场地后部新建的住宅楼为一个家庭住宅，前部则被改建为一家建筑公司的办公室。同时，楼内还提供了当地社区用于举行会议的场所。

该项目诞生于穆特茨/巴塞尔城组织的一项设计竞赛，竞赛要求改建一座位于市中心18世纪的农舍。从外部来看，古老农舍内的住宅和办公区颜色相近，且采用类似颜色的木材、石料和玻璃等材料。农舍内部，办公区的传统构造通过粉刷成白色，已被改造得较为现代化，同新建筑连为一体。浅色与未来在山墙立面增建的窗户使曾经可能昏暗的室内射入阳光，照亮了整个工作区。

新建的Rubido Romero基金会总部无需划出如此严格的界限。房屋原属Rubido Romero家族，最近由Abalo Alonso建筑事务所进

building's upper floors.

The office and residential building in Kirchplatz by Oppenheim Architecture+Design keeps a clear separation between functions for the simple reason that they are used by different groups of people. A family occupies the new residential building at the back of the site, while the existing building at the front has been converted into offices. They are currently occupied by an architectural firm, while the building also provides meeting space for the local community.

The project was borne out of a design competition organized by the city of Muttenz/Basel for the renovation of an eighteenth century farmhouse, in the historic center of the city. Outwardly, the dwelling and the office space in the old farmhouse share a similar material palette, of wood, stone and glass. Inside, the traditional construction of the office space has been modernized and united with the new building, through the use of white throughout. The light color, together with the addition of further windows on the gable facade, makes what was once probably a dim interior into a bright working space.

The new headquarter of the Rubido Romero Foundation does not need to draw such strict boundaries. The house had belonged to the Rubido Romero family and was recently renovated by Abalo Alonso Arquitectos. It is now an office, exhibition and meeting space, run by the former owner's foundation.

The building is a traditional cottage in Lugar de Padin, a collection of six or seven houses in Negreira, a small town near to Santiago de Compostela in Galicia. The architects' intervention was kept to a minimum, so that from the outside, the building retains much of its original appearance. The thick masonry walls needed to be reinforced as part of the works, a new WC was added to the building and the first floor was removed in the entrance area to create a double height space where there had originally been a low ceilinged stable. A line of three relatively small rooms sits between the entrance and the long, thin multi-purpose space. The interior has been painted a uniform white.

Only one person works here full time and their small office is located right in the heart of the building, surrounded on all sides by the more public functions. Enclosed by thick stone walls, it should be possible to shut out the rest of the center's activities, or open up to be at the center of the action. The deep entrance hall, open

照片提供：© Kiyonobu and Nakagame (Shigeo Ogawa)

莫托萨布公寓sYms，向邻近的街区开放，以在相对较小的面积区域内创造一种开放感
Motoazabu Apartment sYms, wide open to the neighborhood to create a feeling of openness in a relatively small area

行修缮。目前，这栋房屋兼顾办公、展览、会议等功能，由原业主基金会进行管理。

这座建筑是位于Lugar de Padin的一栋传统乡间别墅，位于小镇尼格雷亚，有六或七栋房屋，小镇临近加利西亚省圣地亚哥-德孔波斯特拉古城。建筑师的改动尽量控制在最小范围，从外观上几乎看不出改建的痕迹。建筑师加固了厚实的石墙，新建了卫生间，拆除了一层的入口区，以建造一处双层高的空间，这里原是一处低矮带顶的马厩。入口处和狭长的多功能空间之间有3间较小的房间，成线形排列，屋内全部粉刷成白色。

这里只有一名全职工作人员，员工的小办公室位于建筑的中央，周边都是公共空间。厚重的石墙环绕，可以封闭起来，将中心的活动关闭在外，也可以敞开大门，成为活动的中心。进深较长的入口门厅通往外面，与景观有着紧密的联系，也是游客们在多雨的加利西亚躲雨的好去处。

马德里博坦基金会新建的办公室在许多方面与旧建筑都很相似（尽管新办公室的规模更大），这是一座位于旧建筑内的新总部，其布局需要将一系列的公共活动和会议区与行政区完美结合起来。建筑原先是一栋工业大楼，始建于1920年，最初为一家银器店，后被Vincon商店接管。在为新业主重新规划此楼时，MVN建筑事务所提出了两大战略。

战略一为展现建筑历年的变化。因此设计中使用条状的金属色板、玻璃、橡木来与现存的墙砖相得益彰。

战略二是大力增加采光。鉴于此，原有的很多窗户和屋顶天窗都是关闭的，建筑师把它们一一重新打开，让阳光照进曾经密不透风的房间，特别是在这样一个建筑密集的城市中更应如此。

一楼的基金会办公楼是残存的几个房间，由于其与街道的关系，建筑属于内向型，人们必须穿过两栋建筑之间的狭长巷道才能到达入口，因此，博坦办公室的新建入口备受关注。这也是设计中改建最大的部分之一，设计新建了一个全高的中庭，一楼的阳台将上层的私人空间与和街道同水平的公共集聚区和会议区连接起来。环绕采光天窗

to the outside, provides a strong connection to the landscape, while protecting visitors from the frequent rain showers in Galicia. The New Offices of the Botín Foundation in Madrid is similar in many ways, albeit significantly larger – a new headquarters in an old building for an organization that needs to combine a series of public event and meeting spaces with its own administrative spaces. The building was an industrial one – it was built in 1920 to house a silversmith and was later taken over by the Vincon shop. In reimagining it for its new owners, the architects, MVN Architects, had two driving strategies.

Their first strategy was to reveal the changes that the building had gone through over time, so that the design uses a stripped back palette of metal, glass and oak to complement the existing brick. The second was to allow greater amounts of daylight in. To this end, existing windows and roof lights, many of which had been closed up, were reopened, bringing natural light to what must have been a fairly airless space previously, especially in such a densely built part of the city.

The Foundation's offices on the first floor remain at a couple of removes from the world outside. In terms of the relationship to the street, the building is very inward looking. The entrance is approached via a narrow alley between other two buildings, making the Botín Foundation's new entrance space even more dramatic. This is one of the strongest moves in the scheme, the full height atrium, with the balcony on the first floor level linking the private program above to the street level public gathering and meeting functions, is created. This prominent space is emphasized further by dropping the walls around the roof light down to almost meet the balcony, forming the atrium space into a volume in its own right. Trees planted in this space emphasize its vertical quality. This connection between first and second floors is softened by the treatment of the roof light and the solid balcony, which shields the offices from the busyness below. The public areas on the first floor can be separated off from the atrium, and from each other, by a number of transparent and opaque screens. The treatment of quieter spaces for the offices display a similarly light touch. The majority of the space plan is open, with one enclosed meeting room created as a delicate glass box, almost like a piece of furniture. In this way, the architects also quietly draw attention to the difference between the existing industrial building and their

照片提供：©MVN Arquitectos

马德里伯丁基金会的新办公室，试图保留其原始工业特点的精髓
The New Offices of the Botín Foundation in Madrid, attempting to retain the spirit of the original industrial character

的墙壁几乎与阳台连接，使这处区域更加突出，由此形成的中庭自成一体，种植的树木更加凸显了建筑的垂直感。

采光天窗与厚实的阳台使一楼和二楼之间的连接更加柔和，它们将办公室与下层的忙碌相隔离，一楼的公共活动区可用透明或不透明的屏风隔开，使之与中庭分隔，互不干扰。办公楼的较安静区域也同样采光良好。空间的大部分区域都是开放的，只有一间封闭的会议室，像一个精巧的玻璃箱，也像一件家具。设计师用这种方法，悄无声息地抓住了现存的工业建筑和嵌入其中的部分的差别。

FGMF建筑事务所设计的卡萨雷克斯办公室在街区所处的位置十分出挑。这家广告代理机构的新办公室位于圣保罗的帕卡恩布街区，已经部分从20世纪40年代的老建筑外壳中脱颖而出，并且在其后方新扩建了一座新的建筑。建筑师与该广告代理机构联手为这栋老宅建造了一个新立面：将红色与灰色的砾石填充在金属笼子内，背光式的标志为公司的商标，以创造一个醒目的、坚硬的办公室立面。

这栋建筑的室内清晰地划出三片区域：前部右侧是一个可以兼顾接待区与该公司展区的双重功能区；建筑后侧是一个两层楼挑高的工作区，二楼是主管办公室，工作区与接待区中间是会议室，这里是建筑面向公共的一面与后侧办公室相接的地方。

入口大厅是建筑最重要的部分。老建筑自身损害严重，改建成本巨大，建筑师对此采取了反向而行的策略。他们拆除了二层，以建造一处双层高的空间，对现有的楼层表面进行拆毁，刮掉墙体表面的灰泥，以将砌砖裸露出来，其间嵌入白色的交通流线元素，即一个螺旋形楼梯，连接着二层的会议室与一条白色的通道，这条通道引导游客走上一条行程较短但是较为曲折的路线，他们在到达接待处之前会路过工作区的展览空间。小径两侧铺设的花岗岩石块将电缆隐藏起来，远离人们的视线，来访者无捷径可走。

建筑尾部是工作间，为这家广告代理机构提供了工作空间以及私人空间，这一大型体量允许办公室的人们共享同一空间。由预制现浇混凝土构件建造的大型书架依附在一面两层高的墙体上，一个木质楼梯通向主管办公室。而在建筑后方，人们视野不及的地方，是建模室

interventions in it.
FGMF Arquitetos' Casa Rex Office asserts itself more strongly on the street. The new office for this advertising agency in the Pacaembu neighborhood of Sao Paolo has been created partly from the shell of an existing 1940s house, with a large new extension at the back. The architects and the advertising agency collaborated to create the facade design, a pattern of gabions filled with red or gray stones, with the backlit sign featuring the company's logo, to create a bold, tough face for the office.
Once inside, the buildings are clearly divided into three zones. Right at the front, housed in the 1940s construction, is the reception and an exhibition space for the agency's work. At the back of the building is a mostly double height space that serves as the agency's studio. Above this is the director's office and between this working area and the reception, are the meeting rooms where the public face of the building meets the back office.
The entrance "lobby" is the most remarkable part of the building. Having fallen badly into disrepair, a full refurbishment would have been costly. So the architects took the opposite tack, stripping the building back to its fundamentals. They removed the second floor to create a double height space, destroyed the existing floor surface and chipped the plaster from the walls to reveal the brick work. Inserted into this are the pristine white circulation elements – a spiral staircase connecting to the meeting rooms on the second floor level and a white path which leads visitors on a short, but winding route through an exhibition of the practice's work before arriving at the reception desk. Chunks of granite piled either side of the footpath hide cables from view and discourage short cuts. Tucked away at the back, the studio offers the advertising agency the space and privacy to do its work, a large volume, which allows almost the entire office to share the same space. Set against one of the two story high walls is a giant bookshelf constructed from pre-cast concrete elements. A wooden staircase runs up to the director's office, while behind, out of view, is the model area and bathrooms.
The new office of Korean music label Stardom Entertainment, in Seoul was also created from the run-down shell of an existing building. It is located in DokSan, the meat-packing district of the city, an area which has slowly been abandoned by industry as the infrastructure was aged and other locations became more

卡萨雷克斯办公室是一家广告代理机构的新办公室,部分是由20世纪40年代的住宅的外壳改建而成
Casa Rex Office, the new office for advertising agency, created partly from the shell of an existing 1940s house

Stardom娱乐公司办公室保留了现有的混凝土墙体,形成了一个粗糙的表面
Stardom Entertainment Office with the existing concrete walls which has a ragged appearance

和洗手间。

位于韩国首尔的标志性音乐公司Stardom娱乐公司的新办公楼建在一栋残破的老建筑的躯壳之上。老建筑地处首尔市肉类加工区DokSan区,由于基础设施老化,且其他地区的舒适度有所提高,这个市场已慢慢被工业所遗弃,遗留了许多废弃的街区建筑。D·LIM建筑事务所的建筑师们对其进行了改造,尽量保留建筑的历史痕迹,同时使其适应新功能,以体现音乐产业的魅力。

建筑外墙结构合理,修复相对较少,而内墙需要更换一些构件。许多现存混凝土墙壁表面粗糙,但是,建筑师希望在沿着新建的嵌入结构的一侧保留老墙壁,以对建筑历史进行认知。同样,外墙的瓷砖覆层也给予了保留,粉刷成暗灰色。正立面覆有聚碳酸酯材料,标志着它的新用途。

最后的一个项目是日本Mie县松阪SUGAWARADAISUKE建筑事务所设计的森林中的办公室。建筑对室内外之间的关系进行了处理,设计了封闭空间和更开放的内部空间。建筑的体量为一个简单的黑色立方体。然而,室内空间的形式为一个受挤压的房子,从主体量中逐步削减,人们在建筑的边缘均可见这种外形。

室外的景观被邀请入内。巨大的开窗为人们提供了开阔的视野,随着四季的更迭以及天气的变化,各种颜色的植被都能映入人们的眼帘,也因此改变了室内。反射天花板更是将这效应进一步放大,建筑师的设计意图是,人们选择一处地方来工作,不仅仅是基于手头的任务和功能要求,同时也要取决于位于其间的人们的情绪与不同空间的体验质量。

建筑师的确舍弃了某些功能的要求,因此,大部分的办公室是开放式的,但是也有安静的封闭空间——这些空间以白色盒子的形式呈现出来,嵌入到主空间中,主空间的一端为带有玻璃幕墙的会议区,综合来看,建筑提供了各种环境。在开放办公区,你可以安坐自然之中来工作,你也可以寻觅一处适合你工作性质的不寻常之地来办公。

convenient, leaving many of the neighborhood's buildings to fall derelict. D·LIM, the architects who designed the building's transformation, wanted to leave many of the traces of the old building's history, while also making it fit for its new purpose and with a reflection of the glamour of the music industry.

The building's shell turned out to be structurally sound, needing relatively little repair, although certain elements needed to be replaced on the inside. Many of the existing concrete walls have a ragged appearance, but the architects wanted to preserve this alongside the shiny new interventions, in recognition of the building's history. Similarly, the tiles cladding the exterior were kept, painted over in dark gray. Signifying its new use, the front facade has been clad in polycarbonate.

The last project is the Office in Forest by SUGAWARADAISUKE in Matsuzaka, in Mie prefecture in Japan. It also plays with the relationship between interior and exterior, closed off spaces and more open ones within the interior. The building's volume is a simple black cuboid. The indoor space, however, takes the form of an extruded "house" shape, subtracted from the main volume, this shape is visible at the building's sides.

The landscape outside is invited in. Large windows offer generous views out and draw in the varying colors of the landscape with the changing of the seasons and with the weather, to transform the interior accordingly. The reflective ceiling amplifies this effect. The architect's intention is that people should be able to choose a place to work, based not only on the task at hand and its functional demands, but also on the mood of the occupant and the experiential quality of the different spaces.

The architect does cede some things to the requirements of function, so that where the majority of the office plan is open, there are also quieter, enclosed spaces. These take the form of white boxes inserted into the main space, plus a glass-walled meeting area at one end. Taken together, they offer a variety of different environments. While in the open part of the office you get to choose a place to work surrounded by nature, it is also possible to find a different place, suited to the nature of your work. *Alison Killing*

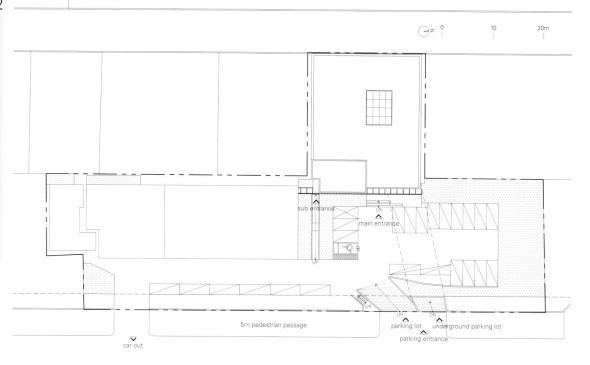

Stardom娱乐公司办公室的重建
D·Lim Architects

肉类加工仓库重建

　　始建于1973年，地处韩国首尔市秃山洞肉类加工区，秃山洞是首尔主要肉类加工区之一，由于基础设施老化，交通不便，首尔大多数的传统市场已慢慢衰败，秃山洞也处于严重的衰败中，该地区的一个废弃仓库被D·Lim建筑事务所改建成为Stardom娱乐办公室。

　　D·Lim建筑事务所从美学和经济方面对建筑进行了改造。现有的外墙几乎未动，而受损部分则选择性地进行加固，室外覆层在过去经过多次的翻修已经显得破旧不堪。外墙的旧瓷砖被涂成暗灰色，以保留建筑原貌和这一地区的历史痕迹。D·Lim建筑事务所只在临主街的正面增建了一处立面，透明的聚碳酸酯嵌板铺在五彩缤纷的涂鸦墙与暗灰色的墙体之下。墙上的涂鸦艺术是Stardom娱乐办公室创始人、现已退休的韩国流行音乐家Cho PD身穿夹克的形象。透明外墙上的涂鸦是Stardom品牌形象的象征。走进一楼宽敞的休息室，受训想成为明星的学员会看到地板上的带状线。入口处是这条线的起点，而在其终点，和煦的阳光穿过庭院照射进来。楼梯位于原先汽车电梯的位置，现在已经不需要汽车电梯了。建筑内部大部分的混凝土表面都未做装饰，裸露在外，显示建筑的本质纹理和岁月的痕迹。

　　犹如曼哈顿的肉类加工区在城市贫民窟重建为最时尚的社区，D·Lim建筑事务所期望Stardom办公室能为韩国肉类加工区带来全新的改变。

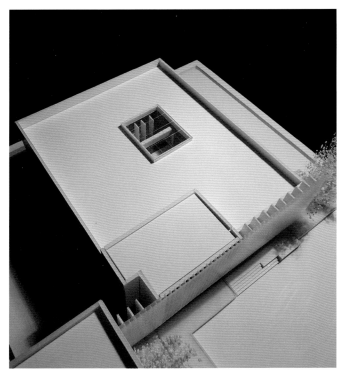
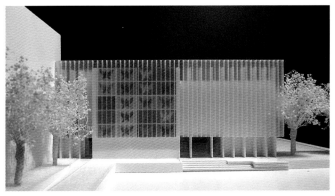

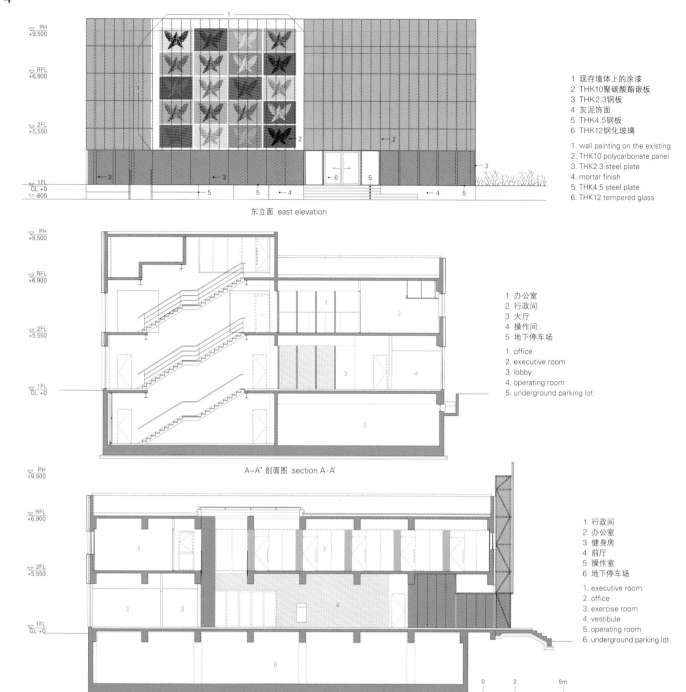

Stardom Entertainment Office Remodelling

Regeneration of the Meatpacking Warehouse

The building is located in DokSan-dong, the meatpacking district of Seoul, Korea. Founded in 1973, DokSan-dong is one of the major meatpacking districts in Seoul. However, as most of the traditional market in Seoul has been declined because of aging facilities and inconvenient transportation, DokSan-dong is also in heavy decline. An abandoned warehouse in the area was remodeled to the Stardom Entertainment Office by D·Lim Architects.

D·Lim renovated the building delicately and economically. The existing outer shell of the building is almost intact and vulnerable structures are selectively reinforced. The exterior cladding looked like rags because it was repaired several times during its life. But, the exterior old tiles are coated only with the dark gray paints in order to leave the trace of the building as well as the district. D·Lim has only added a new facade on the front facing the main street. The translucent polycarbonate panels are padded on the colorful graffiti wall and the dark-grey wall. The graffiti art on the wall is the retirement album jacket image of Korean hip pop musician Cho PD, a founder of the Stardom Entertainment Office. The translucent outer shell on graffiti images symbolizes its brand image. Stepping into the spacious lounge on the first floor, the trainees who want to be a star would see the lace lines on the floor. The entrance is a starting line for them and at the end of the lines, the genial sunshine glares down through a courtyard from the sky light. The staircase is moved to the space used for the car lift which is unnecessary now. Most of the interior concrete finishes are exposed as they are, showing their genuine textures and trace of time.

As if the meatpacking district in Manhattan has been resurged to the most fashionable neighborhood from urban slums, D·Lim Architects expects the Stardom office to lead a new change of the Korean meatpacking district.

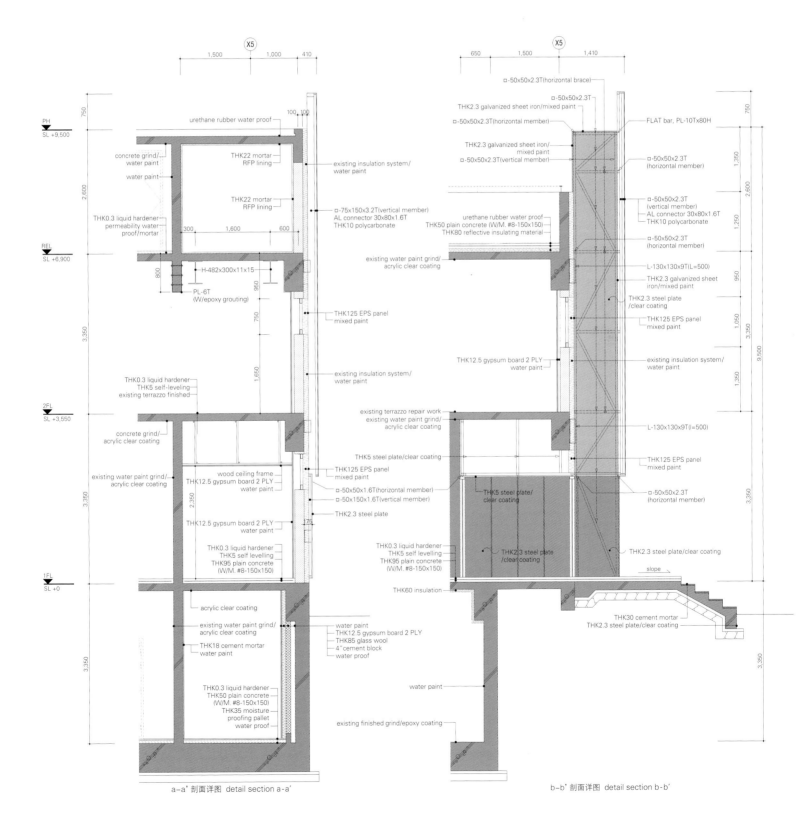

a-a' 剖面详图 detail section a-a'

b-b' 剖面详图 detail section b-b'

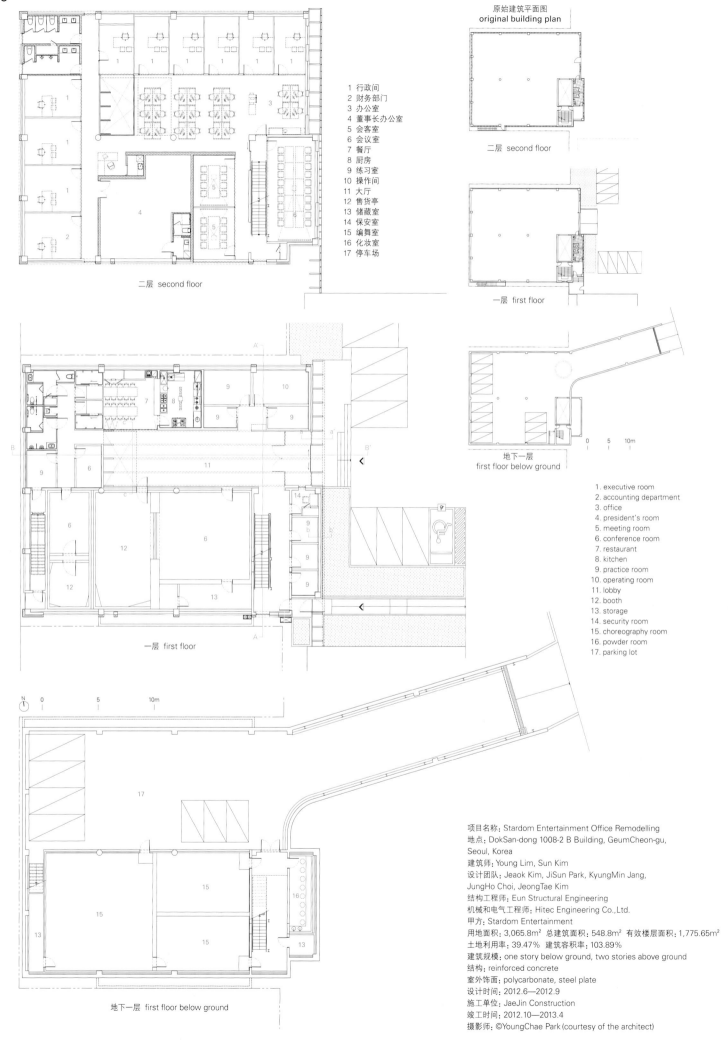

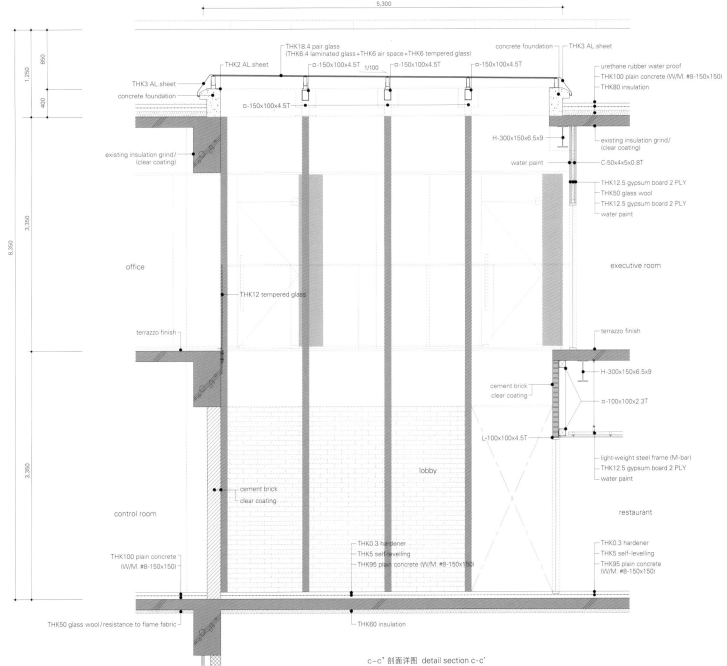

c-c' 剖面详图 detail section c-c'

卡萨雷克斯办公室
FGMF Arquitectos

项目名称: Casa Rex Offices
地点: São Paulo, Brazil
作者: Fernando Forte, Lourenço Gimenes, Rodrigo Marcondes Ferraz
协调员: Ana Paula Barbosa, Marilia Caetano, Sonia Gouveia
建筑师: Bruno Araujo, Bruno Milan, Marina Almeida
学员/实习生: Claudia Bicudo, Felipe Bueno, Gabriel Mota, Gabriela Eberhardt, Mirella Fochi, Rafaela Arantes, Rodrigo de Moura, Talita Silva
甲方: Casa Rex
用地面积: 631m²
有效楼层面积: 603m²
竣工时间: 2012
摄影师: ©Rafael Netto (courtesy of the architect)

卡萨雷克斯是巴西重要的设计公司,它委托建筑师对其新地点进行一次改造,项目的最初目标是为办公室空间创造一个全新的身份,这在之前的快速扩建中是不存在的。

业主方选择了位于帕卡恩布的一栋建于20世纪40年代的住宅来作为新办公室的场地,这栋公寓在过去的几十年里经过了若干次的改造,且几乎掩盖了建筑原来的样子。因此,从一开始,这座建筑的功能便被分为三个部分:一个可以兼顾接待来客与展示该办公室设计的项目的会客室;办公室区域,与会客区完全隔离,所有的员工都工作在这处开放的空间里;最后的一部分是位于前面位置的室外区域,这处区域的进出受到了严格的控制。

因为要考虑预算限制的问题,所以建筑师在设计的过程中充分考虑了一些不同寻常的资源:即很多废弃的材料以及一些基础设施构件的利用。

建筑师在立面中堆积了石头,外面覆有笼子,这是一种常用于道路挡土墙的材料、建筑师与雷克斯团队联手为这座建筑制作了一个特殊的标记,即采用红色的砂岩与灰色的大块砾石,以形成一个在视觉上令人震撼的立面。其中的一个模块加以突出,并且覆上一块板,以凸显办公室的名字。

在接待区和会客区,建筑师拆除大部分结构:他们建造了一个两层高的天花板,拆除了地板,刮掉了原来墙体上的灰泥,使其以考古建筑的形式呈现出来。而在进行拆除的室内空间,建筑师修建了一条洁净纯白的小路,引导着访客通向两个会议室,即位于这些废墟内部的全白色体量。空间的其他部分也全部被砂砾立面所覆盖,办公室的展厅则会嵌入到中间空间,作为废墟之间的展廊。

工作室区域被设计成二层高,天花板占据了整个空间的一半以上。原有结构的一部分二层天花板区域被改建为公司董事长的一间大办公室。在两层高的天花板上,一个云状的照明设备照亮了这处区域,此外,它还发挥着划定空间的作用。在二层高的后墙后面,建筑师设计了浴室、财务区以及建模室,此外,他们还利用堆积的、预先成型的混凝土体块(一般用于疏导水流、设计立面)制成一个面积为70m²的书架,建筑师还把几个木质楼梯嵌在大书架中,除了其本身作为模块的功能外,它还发挥着连接主管办公室和其他部门的作用。

重型建筑材料以全新的方式被利用,建筑师用拆除的瓦片和平板建成了一处独特的空间,与现有的办公室完全不同,这处空间被看做是画廊与企业空间的结合体,与常规的公司略显不同。

Casa Rex Office

When Casa Rex, an important Brazilian design office, came to the architects looking for a reform of its new seat, the initial goal was to create a new identity for the physical space of the office, which didn't exist before because of its rapid expansion.

A former residence in Pacaembu, built in the 1940's, was chosen. The property had several reforms poorly made through the decades that hid almost entirely the building's original architecture. Since the beginning, the clients' program was divided into three parts: the meeting area, where there would be the reception and a place exhibiting the office's projects; the studio's area, completely separated from the first one, where everybody would work together in an open space; and, finally, the external front area, with restricted and controlled accesses.

Having issues such as the limited budget in mind, the architects decided to use some unusual resources while designing the building: a lot of demolition and the usage of some infrastructure pieces.

In the facade they piled stone inside the gabions, a material used as an earth retaining on the roads and the like; together with the Rex team they made a specific pagination where they used red

详图1 detail 1　　0　1　2m

metalic structure
gabion module

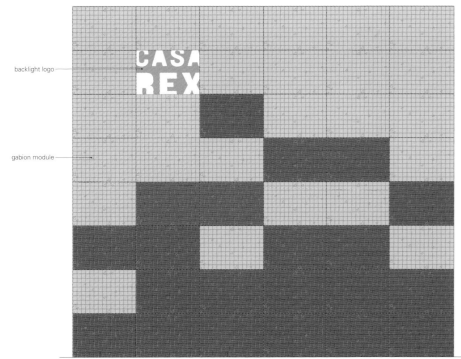

南立面 south elevation

backlight logo
gabion module

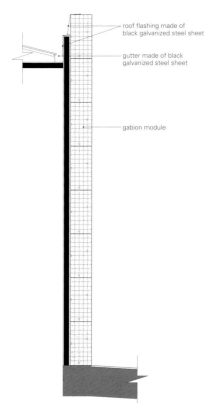

roof flashing made of black galvanized steel sheet

gutter made of black galvanized steel sheet

gabion module

a-a' 剖面详图 detail section a-a'

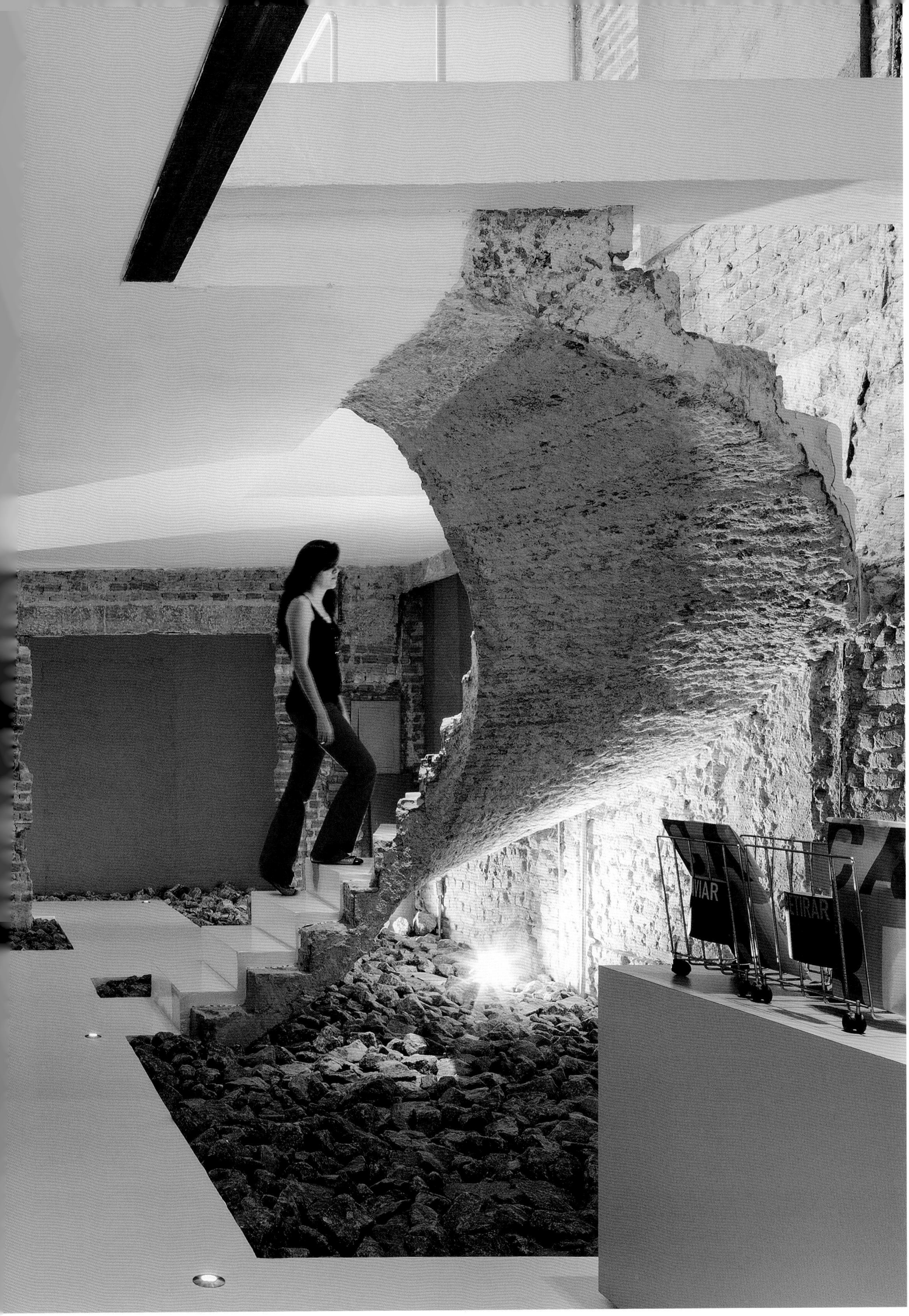

sandstone and big gray gravel to form a visually remarkable facade. One of these modules was suppressed and they put a plate with negative cuttings indicating the office.

In the reception and meeting area, the architects made an extensive demolition: they created a two-story high ceiling, destroyed the floor and removed every plaster of the original walls, in a form of "architectural archeology". Inside this space with a demolished aspect, they created a pure, clean, white path leading the visitor to the two meeting rooms – entirely white blocks inside those ruins. The rest of the space was filled with the same gravel of the facade, and it is in that "inter-space" where the exhibition of the office's projects will be mounted, almost as a gallery amidst the building's ruins.

As for the studio's area, it was designed as a two-story high ceiling – which was built – occupying a little more than a half of the entire space, and a section of normal ceiling is original from the construction, where the upstairs act as a large room for the company's president. On the two-story high ceiling, a "cloud" of luminaires helps to lighten the place, in addition to delimiting spaces. On the two-story high ceiling back wall, behind which the architects found the bathrooms, the financial area and the models area, they made a 70m² bookcase with piled pre-shaped concrete pieces that are normally used for channeling streams, consigning the facade. Into this big bookcase, they have incorporated some wooden stairs, in addition to the modules themselves, in order to reach the footbridge connecting the director's office to the rest of the agency.

Using heavy building materials in an innovative way as well as demolishing tiles and slabs creates a unique space, very different from the existing offices. It is almost as a mix of a gallery and a corporative space, something different for an unusual company.

FGMF Arquitectos

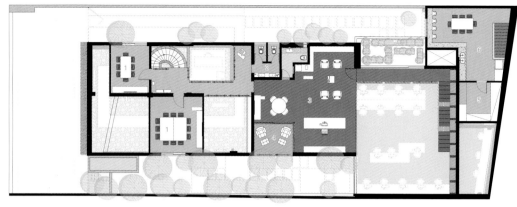

1 会议室	1. meeting room
2 卫生间	2. toilet
3 主管办公室	3. director office
4 露台	4. terrace
5 储藏室	5. storage
6 厨房	6. kitchen

二层 second floor

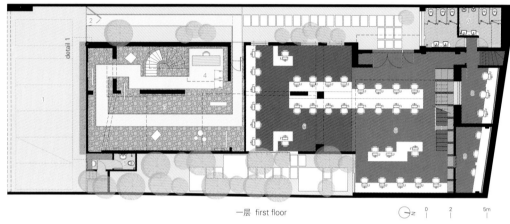

1 停车场	1. parking
2 主入口	2. main entrance
3 展厅	3. exhibition
4 接待处	4. reception
5 保安室	5. security room
6 工作室	6. studio
7 卫生间	7. toilet
8 行政办公室	8. administration room
9 实物模型室	9. mock-up room

一层 first floor

森林中的办公室
Aquaplannet办公总部
SUGAWARADAISUKE

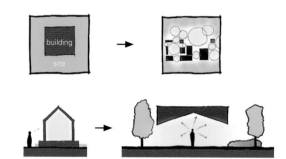

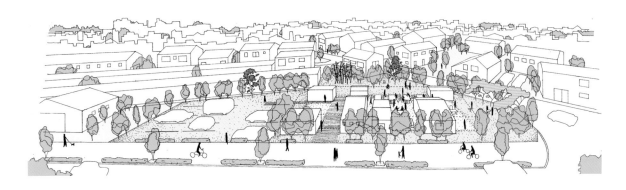

这里是Aquaplannet在松阪的总部,坐落于日本Mie县内。这项工程的目的不在于凸显人们所能看到的这座建筑物的雕刻般外貌所产生的价值,而在于显示考虑到人体感受的环境所产生的价值。

包括员工具体活动范围在内的这一大型体量被规划在种有植被的场地内。有着不同宽度、高度和深度的场地建在同一体量内。通过设计室内空间和作为同等空间元素的花园,人们透过窗户观看风景,并且室内外之间的联系也建立起来。体量内的大型反射天花板将室内元素与外部的各种色彩和光源较好地融为一体——随着季节、气候和时间的改变而改变。

花园的多样性使其从一个体量转换为一个有着不同场地组的异构空间。人们在这一体量内的体验也会随着季节、天气和活动的改变而产生巨大的变化。

这一工程提出了办公室设计的一个新方法,它使得人们在选择工作场所时不仅仅将其功能或是类型作为依据,同时也要考虑到该空间体验的品质及其通过反射的自然对于人体所产生的影响。

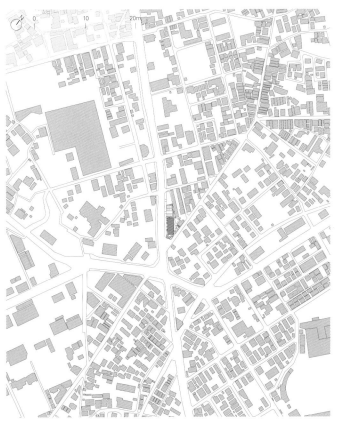

Office in Forest
Aquaplannet Headquarters Building

This is a head office of Aquaplannet in Matsuzaka, Mie prefecture in Japan. The project's target is not on the value of the view in respect to sculptural appearance of the building, but rather on the value of the environmental view with body experience.

The large volume, including working activities, is planned in the site with planted trees. Various places are created in one volume with different widths, heights and depths, sceneries from a window and relationship between in-outside by designing inside spaces and gardens equivalently as spatial elements are built. The large reflective ceiling on the volume integrates inside elements with various colors and lights from outside – changing with different seasons, climates and time.

Diversity of the garden transforms from one volume into a heterogeneous space with groups of different places. Body experience in the volume changes dramatically according to seasons, weathers and activities.

This project proposes a new approach to office design, which enables the architects to choose working places not only by function or work typology, but also by the experiential quality of a space and its effect on the human body through reflected nature.

sky framing

90 degrees framing

tunnel framing

horizontal framing

项目名称：Office in Forest / Aquaplannet Headquarters Building
地点：Matsuzaka City, Mie Pref, Japan
建筑师：SUGAWARADAISUKE
设计发展：SUGAWARADAISUKE, Kitamuragumi Corporation
用地面积：1,193.30m² 总建筑面积：356.44m² 有效楼层面积：348.01m²
建筑规模：one story
结构：steel frame
设计时间：2010.8—2011.12 施工时间：2011.6—12
摄影师：©Nacasa & Partners

西北立面 north-west elevation

1 办公室 2 董事长办公室
1. office 2. president room
A-A' 剖面图 section A-A'

1 多功能室 2 入口大厅 3 办公室 4 董事长办公室 5 设计区域
1. multipurpose room 2. entrance hall 3. office 4. president room 5. design area
B-B' 剖面图 section B-B'

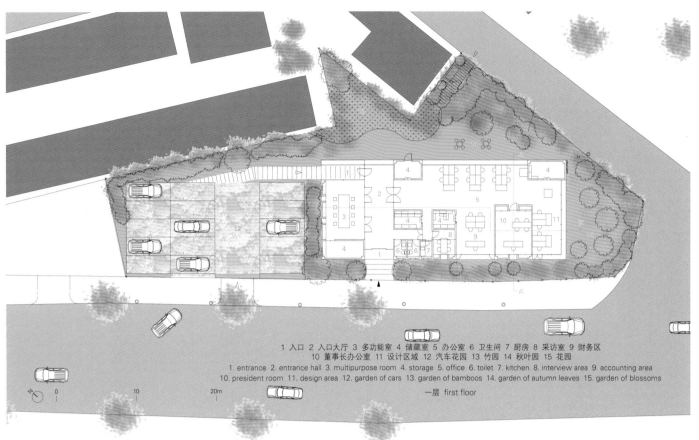

1 入口 2 入口大厅 3 多功能室 4 储藏室 5 办公室 6 卫生间 7 厨房 8 采访室 9 财务区
10 董事长办公室 11 设计区域 12 汽车花园 13 竹园 14 秋叶园 15 花园
1. entrance 2. entrance hall 3. multipurpose room 4. storage 5. office 6. toilet 7. kitchen 8. interview area 9. accounting area
10. president room 11. design area 12. garden of cars 13. garden of bamboos 14. garden of autumn leaves 15. garden of blossoms

一层 first floor

微工作·微空间 | Minor Works

Rubido Romero基金会
Abalo Alonso Arquitectos

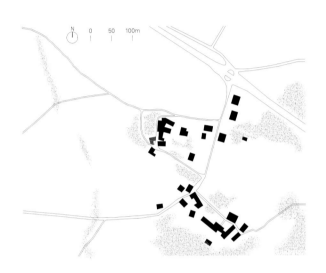

建筑师曾被要求将加利西亚农村地区即将作为基金会总部的大楼进行翻修，那是一个规模较小的传统式建筑。建筑师将对该项工程采取一定的特殊关照；墙壁会被加固；瓦屋顶会被撤换，同时，他们会安装一个非常适用于周边环境的浴室。房屋内部会被翻修并且粉刷成白色。部分墙体将会被镶嵌板，以替代之前的木结构。嵌入其中的房梁将会被用来支撑照明设备，因此得以再次使用。花岗岩地板和黑色钢架的收尾工作将会是整项工程的尾声。

在被赋予了新型的公用功能之后，原有的两层高的天花板高度就显得有些不合适了。由于一楼曾被作为畜舍使用，因此那里的天花板相当低。鉴于此，入口区将会有4m的高度。对于其新增功能来说，这一高度或许还会更高。空调设备将会被放置在浴室上部，通风烟囱会和照亮整个区域的天窗相连。在其他的房间里，天花板高度将会被缩减，以此来与自然地形的斜度保持一致。

通往室外的大厅以及与其毗邻的储存室将会保留下来。这一举措对于像加利西亚这样的多雨气候地区来说具有很大的实用价值。由于受到了较少的且相对简单的外界干预，在建筑物其他区域所能预见到的这一灵活性，使其自身在今后将成为一个展示乡村活力的小中心；在这里，人们可以进行一些诸如公开讨论、展览、休闲活动以及相关事宜的咨询等小型活动。

Rubido Romero Foundation

The architects have been asked to renovate a small traditional building in rural Galicia which is to be used as the foundation's headquarters. Particular care will be taken in the project; walls will be reinforced; the tiled roof will be replaced and a bathroom adapted to its surroundings will be installed. The interior will be repaired and painted white; some walls will be panelled and the woodwork will be replaced. The intervening beams will be reused as a support for the lighting. Granite flooring and finishing touches of black steel will complete the work.

Given its new public use, the ceiling height of the original two floors is inadequate being that the first floor was used as a stable and its ceiling is particularly low. For this reason, the access area will have a height of four meters, perhaps this will be more in accordance with its newly endowed function. The air-conditioning unit could be placed over the bathroom and the ventilation chimney joins with the skylight which illuminates the area. In other rooms, the ceiling height would be reduced, conforming to the slope of the natural terrain.

The hall leading out to the exterior, and its adjoining storage room, will be kept. This is practical in a rainy climate such as that of Galicia. The flexibility foreseen in the rest of the building, with simple and limited intervention, enables its future use as a small center of rural dynamism: public talks, exhibitions, leisure activities, counseling on related matters, etc.. *Abalo Alonso Arquitectos*

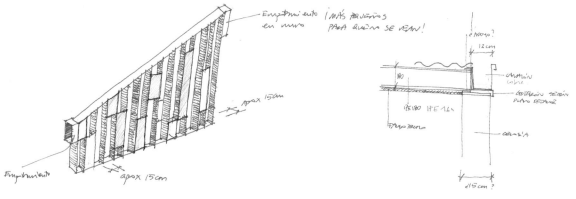

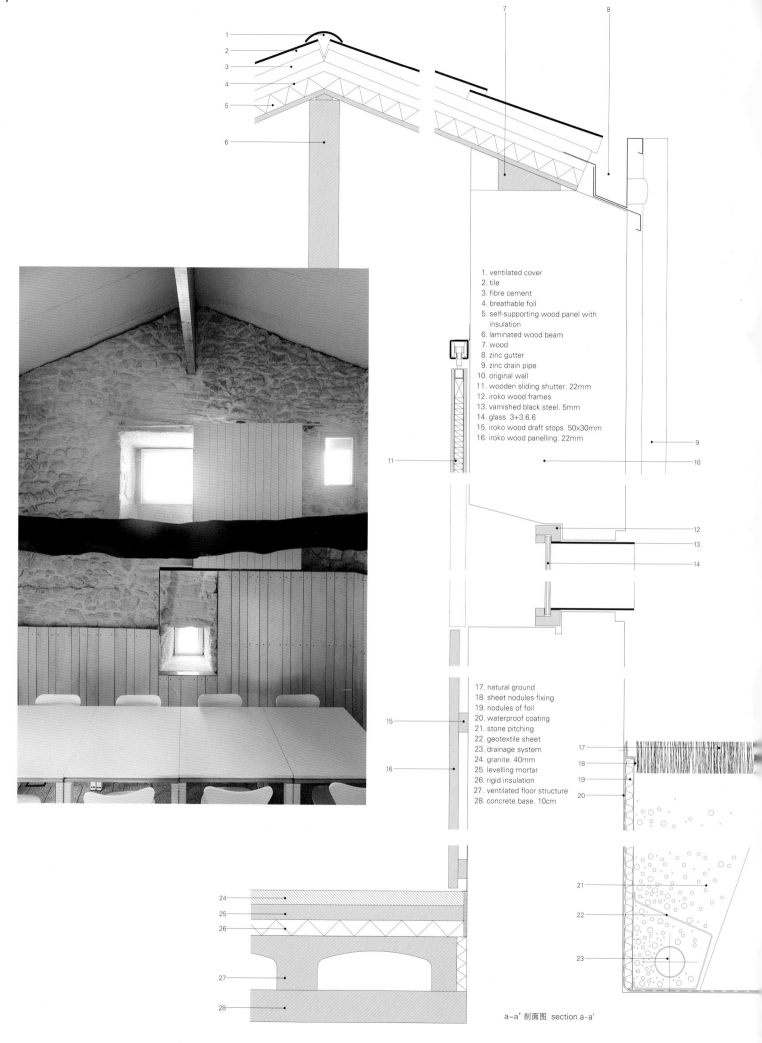

项目名称：Fundación Rubido Romero
地点：Lugar de Padín. Negreira, A Coruña, Spain
建筑师：Elizabeth Abalo, Gonzalo Alonso
合作者：Berta Peleteiro
结构：Carlos Bóveda
设备：Inaec ingeniería
技术建筑师：Francisco González Varela
发起人：Fundación Rubido Romero
总建筑面积：175m²
造价：EUR 152,755.35
竣工时间：2012
摄影师：©Héctor Santos-Díez (BIS Images)

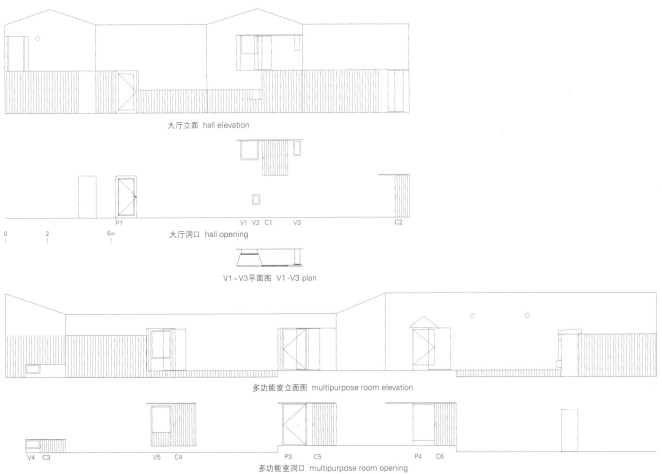

大厅立面 hall elevation

大厅洞口 hall opening

V1~V3平面图 V1-V3 plan

多功能室立面图 multipurpose room elevation

多功能室洞口 multipurpose room opening

1 入口
2 储藏室
3 大厅
4 多功能室
5 办公室
6 机械服务间
7 洗手间
8 走廊

1. entrance
2. storage room
3. hall
4. multipurpose room
5. office
6. mechanical services
7. wc
8. corridor

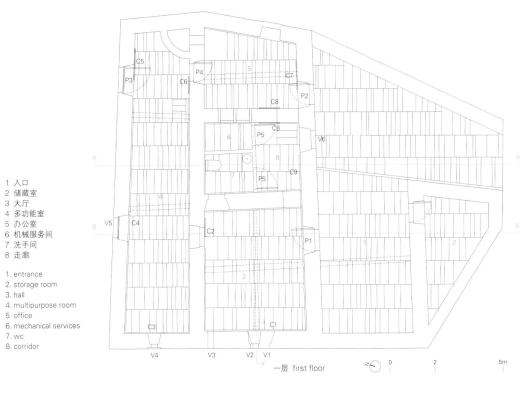

一层 first floor

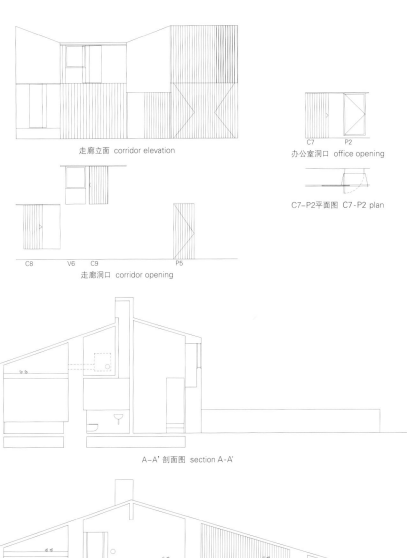

走廊立面 corridor elevation

办公室洞口 office opening

C7-P2平面图 C7-P2 plan

走廊洞口 corridor opening

A-A' 剖面图 section A-A'

B-B' 剖面图 section B-B'

马德里博坦基金会的新办公室
MVN Arquitectos

博坦基金会在马德里选取了由建筑师贡萨洛·阿瓜多早在20世纪20年代设计的一栋工业建筑来作为自己办公室的新址。在之前的岁月里,这座建筑物曾是Luis Espuñes银器工作室,近些年则成为马德里的Vinçon店。

这栋楼的特别之处在于,它给了建筑师一个创造参考点的机会。考虑到这一点,为了与基金会新项目的哲学思想保持一致,新项目尝试保留住其原始工业特点的精髓,而这,正是开发潜力的驱动力所在。

这一建筑概念的主要目的是使日光能够再次地渗透到整座楼内,这座大楼在上次被利用时一直处于阴暗的状态。不仅仅窗口和天窗得以重新利用,内部结构也发生了变化,来营造一个双高度的中庭,作为主大厅。直射的日光以及自然生长的植物让这个聚会地点更加具有了一定的特色和个性。

这一项目旨在通过展现原始的钢构件、砖砌体来体现这座建筑物的历史变迁;以往种种的改变与材质为橡木、钢构件以及玻璃的新结构饰面形成对比。一楼建成一个可供人们进行各类社会活动的、灵活的、模块式的开放性空间。这里有两处划分开的活动区域,其中一个区是不透光的,而另一处区域是透明的,有着四种可能的安置场所。人们可以根据每种活动的不同需求将这一空间按照各种可能的组合方式进行简易的安装。原木在地板和天花板中的运用为这片区域增添了一丝暖意。

二楼将由基金会的管理部门使用,那里设有一个用于会议的私人区域,以及位于原有的屋顶结构和新安置的照明设备之下的宽敞、明亮的开放区。两个分区都被合理地安置在新采光井的周边,在中庭处悬挂着的吊灯的辉映下愈发完美。整个规划中所要求的唯一一个封闭式空间,是一个采用了最小的框架结构的箱式玻璃房。它就像是一件家具一样,不需要任何外界物体的支撑,因而不会与整个空间产生冲突。

The New Offices of the Botín Foundation in Madrid

The Botín Foundation has established its new offices in Madrid by choosing a 1920s industrial building by the architect Gonzalo Aguado, which was for years the Silversmith workshop of Luis Espuñes and more recently the Vinçon shop in Madrid.

The peculiarity of this building has given the architects the opportunity to create a reference point. With this in mind, the new project attempts to retain the spirit of the original industrial character, in keeping with the same philosophy of the Foundation, which is a driving force for developing talent.

The principal objective of the architectural concept is that once again, natural light can enter the whole building, which has remained in the dark since its last use. Not only the infilled windows

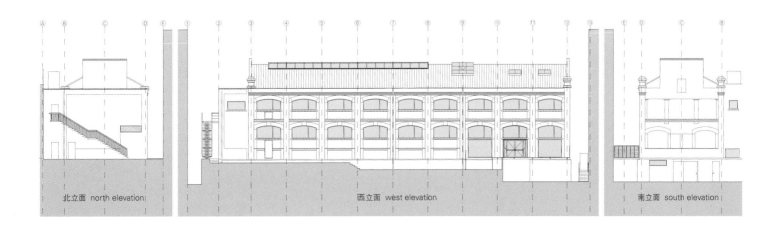

北立面 north elevation　　　西立面 west elevation　　　南立面 south elevation

and skylights have been reopened, but also the internal structure has been altered to create a double-height atrium for the use as the main lobby. The direct daylight and natural vegetation impose character and personality to this meeting place.

The project aims to reveal the historical changes of the building by exposing the original steel and brickwork; the various alterations in the past are in contrast to the new construction work of which finishes are mainly oak, steel and glass. The first floor is intended for public activities with a flexible but modular, clear open space. There are two movable partitions, one opaque and the other transparent which has four possible positions. Depending on the different requirements of each activity, the space can be easily adapted by the various combinations available. The use of natural wood on the floor and ceiling adds warmth to this area.

The second floor is to be used by the management of the Foundation with a private area for meetings and the spacious, luminous open area, under the original roof structure and the new skylights. The two areas are arranged around the new light well and crowned by a lantern over the atrium. The solution for the only enclosed space as required by the program, is a glass box with minimal framework. Free-standing, like a piece of furniture, it does not interfere with the space as a whole.

项目名称：The New Offices of the Botín Foundation in Madrid
地点：Madrid, Spain
建筑师：Diego Varela de Ugarte, Emilio Medina Garcia
项目经理：Santander Global Facilities
合作者：Alfonso García del Rey, Laura Sánchez, Maria Pascual, Alicia Castilla
技术建筑师：Maria Lamela Martin
室内设计顾问：Juan Luis Líbano
顾问：Ingenor, Luis Vallejo
承包商：Ferrovial
甲方：Botín Foundation
有效楼层面积：1,541.95m²
造价：EUR 1,784,000
竣工时间：2012
摄影师：©Alfonso Quiroga (courtesy of the architect) (except as noted)

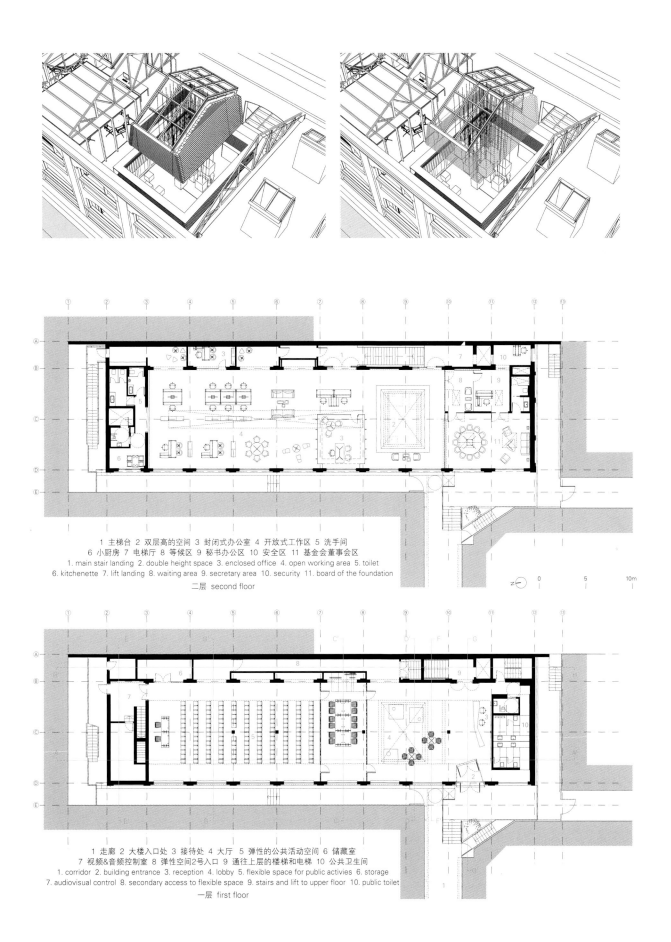

1 主梯台 2 双层高的空间 3 封闭式办公室 4 开放式工作区 5 洗手间
6 小厨房 7 电梯厅 8 等候区 9 秘书办公区 10 安全区 11 基金会董事会区
1. main stair landing 2. double height space 3. enclosed office 4. open working area 5. toilet
6. kitchenette 7. lift landing 8. waiting area 9. secretary area 10. security 11. board of the foundation

二层 second floor

1 走廊 2 大楼入口处 3 接待处 4 大厅 5 弹性的公共活动空间 6 储藏室
7 视频&音频控制室 8 弹性空间2号入口 9 通往上层的楼梯和电梯 10 公共卫生间
1. corridor 2. building entrance 3. reception 4. lobby 5. flexible space for public activies 6. storage
7. audiovisual control 8. secondary access to flexible space 9. stairs and lift to upper floor 10. public toilet

一层 first floor

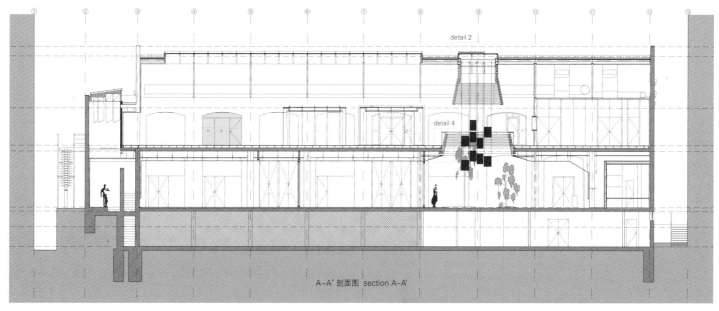

A-A' 剖面图 section A-A'

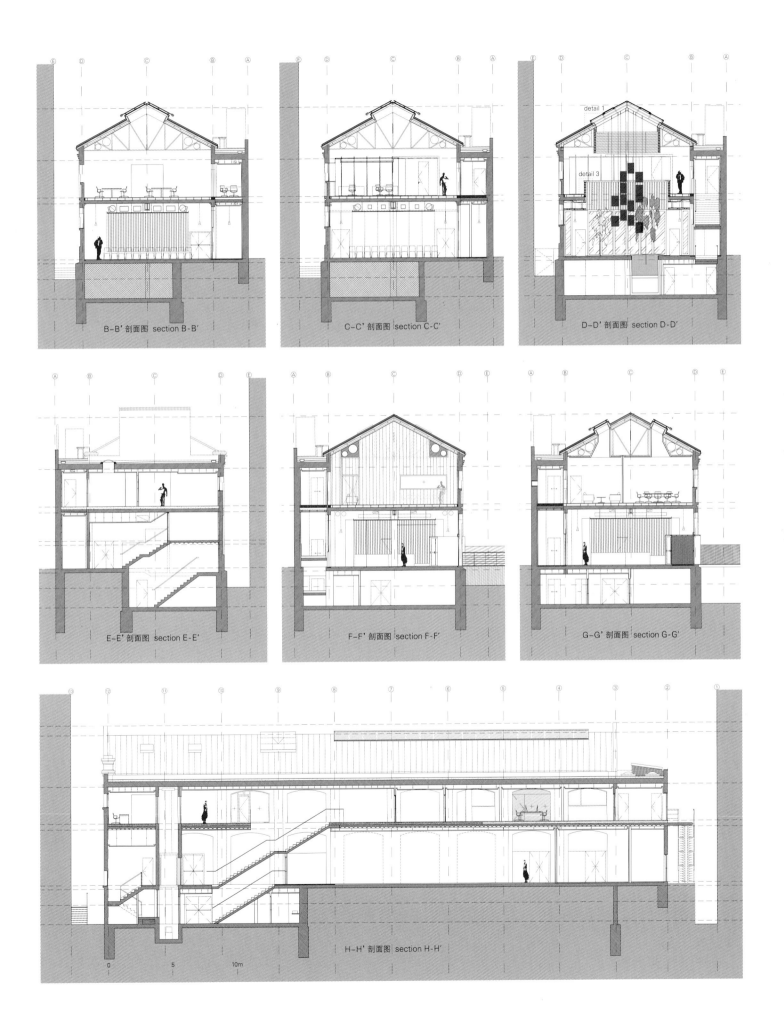

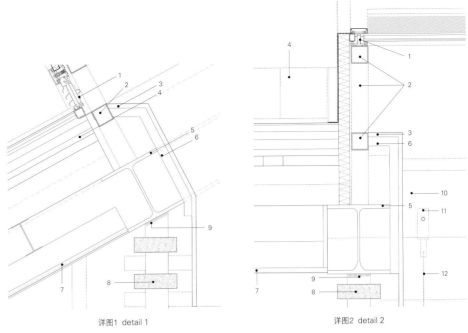

详图1 detail 1　　　　　　　　　　　　详图2 detail 2

1. sistema de carpinteria con aireador inferior 2. ø60.5 3. placa de carton yeso 15mm 4. cubierta de zinc 5. hea-220 6. omega de acero galvanizado e=0.55mm 7. placa de carton-yeso 1.5mm 8. lama de madera 15x5cm 9. pletina soldada 10. plantina acero 21x0.15cm 11. plentina acero e=3mm 12. cable acero

1. vidrio laminar 5+5 translucido mate 2. remate de madera 50x5cm 3. lama de madera 15x5cm 4. lama de madera 116x15 5. soporte metalico 6. pletina de acero e=3mm 7. IPE 400 8. mortero perlifoc 9. perfil ld 80.60.6

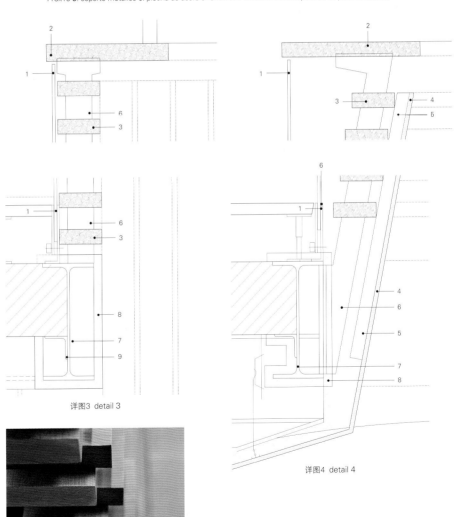

详图3 detail 3　　　　　　　　　　　　详图4 detail 4

Kirchplatz办公室和住宅
Oppenheim Architecture+Design

东立面 east elevation

南立面 south elevation

该项具有适应性的再利用工程源于巴塞尔Muttenz城所首创的一项设计竞赛。这一设计建立在对位于该城市历史中心一所古旧农舍进行翻新的基础之上。这座农舍最初是在1743年建造的。如今,改建后的农舍成为一所建筑设计公司的办公室所在地。这里为社区会议提供了相应的场所,同时也是通往一处明显的毗邻崭新私人居住区的连接区域。

这一新型设计旨在对现存的古旧农舍建筑所具有的传统特色及室内进行一次全新的解读。它是通过增加用于日光照射的洞口和在房屋内部使用纯白色饰面而实现的,且与木质纹理并置,形成对比。各个空间通过这种方式打开,彼此也能够重叠以及合并在一起。

这一可持续性思考包括使用现行的迷你能源(能效)工程标准来对一座高能效的建筑物开展维护。其他维修手段还有诸如使用太阳能屋面板,使用再生性木料来作为建筑物立面材料的可持续性选择,以及对可能的现有建筑元素进行整修。

本工程也包括一个新式独立房屋的设计。那里紧挨着先前可再利用的被改造成为办公区域的古旧农舍。这一优雅的现代住宅结构与古旧的建筑相互辉映。现代与古典在此共享着材质与色彩的共通性,而这二者的结合也给人们的感官带来了惊喜,从而形成了截然不同的表达方式。这栋三层住宅的顶楼是主卧和客房;厨房、餐厅和客厅位于一层;儿童卧室则被安排在地下,并且设有一个通向一楼的斜坡式室外后院露台。

Kirchplatz Office and Residence

The design of this adaptive re-use project was born initially out of a design competition initiated by the City of Muttenz, Basel. The design was based on the renovation of an historic farmhouse situated within the historic center core of the city. The original farmhouse was constructed in 1743. Today the converted farmhouse serves as an office for an architectural design company, provides community meeting space, and serves as a compelling link to a new, adjacent private residence.

The new design aimed to provide a fresh interpretation to the existing traditional features of the historic farmhouse building and it's interior. This is achieved by creating new openings for natural

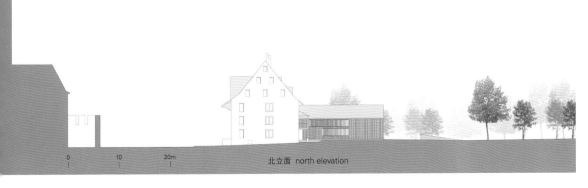
北立面 north elevation

西立面 west elevation

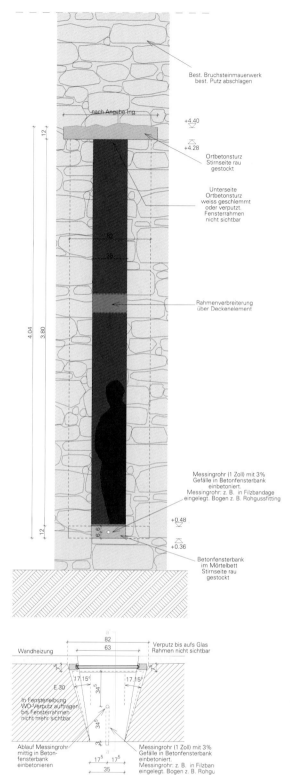

南侧窗户详图 south side window detail

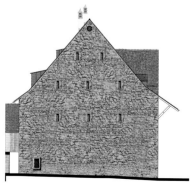

南立面_翻修前 south elevation_before

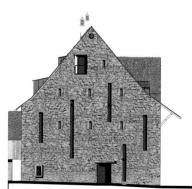

南立面_翻修后 south elevation_after

daylight and by using a crisp white finish in the interiors, which juxtapose against the texture of the old wood and through the way, the spaces open up, overlap, and merge together with one another.

The sustainability considerations included maintaining an energy-efficient building through the use of current Minergie (energy efficiency) construction standards, solar roof panels, a sustainable choice of materials such as reclaimed wood used for the facade, and the restoration of existing architectural elements where possible.

The project also included the design of a new single family house adjacent to the re-used historic farmhouse that was converted into the office. This elegant contemporary residential structure juxtaposes with the historic building. The new and old share commonalities of materials and colors, yet have distinctly different expressions with the interplay of modern and historic delighting of the senses. The 3-floor house is organized with the master bedroom and guest bedroom on the top floor; the kitchen, dining and living spaces on the ground level; and the children's bedrooms below ground with a ramped outdoor backyard terrace leading up to the ground level.

a-a' 剖面图 section a-a'

项目名称：Kirchplatz Office & Residence
地点：Muttenz, Basel, Switzerland
建筑师：Oppenheim Architecture+Design
室内设计师：Oppenheim Architecture+Design
协作者：Huesler Architekten
用地面积：1,300m²
有效楼层面积：办公室面积：280m²，住宅房屋面积：300m²
公寓1_250m²，公寓2_240m²
竣工时间：2012
摄影师：©Børje Müller (courtesy of the architect)

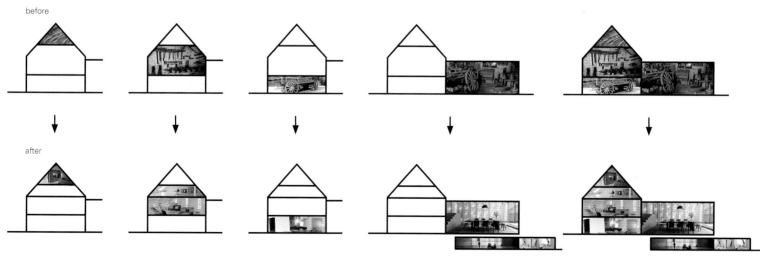

before

after

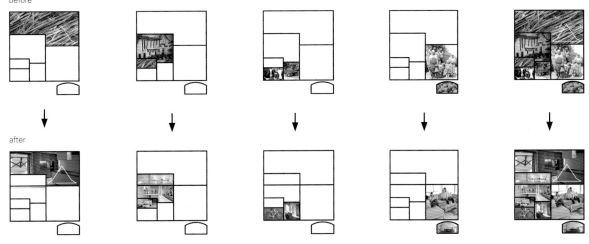

before

after

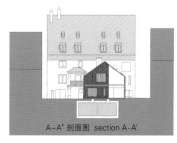
A-A' 剖面图 section A-A'

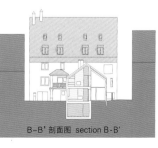
B-B' 剖面图 section B-B'

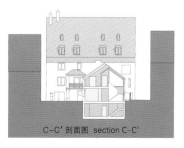
C-C' 剖面图 section C-C'

D-D' 剖面图 section D-D'

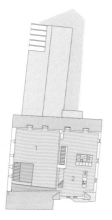

1 办公室 2 公寓-2
1. office 2. apartment-2

三层 third floor

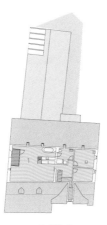

1 公寓-2
1. apartment-2

四层 fourth floor

屋顶 roof

1 会议室 2 储藏室 3 技术室 4 盥洗室
5 休息室 6 桑拿房 7 洗手间 8 卧室
1. meeting room 2. storage 3. technical room 4. washing room
5. lounge 6. sauna 7. restroom 8. bedroom

地下一层 first floor below ground

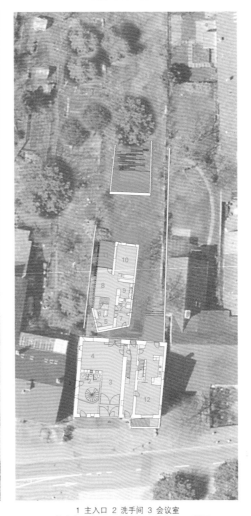

1 主入口 2 洗手间 3 会议室
4 休息室 5 入口 6 更衣室 7 会客室 8 餐厅
9 厨房 10 起居室 11 露台 12 公寓-1
1. main entrance 2. restroom 3. meeting room
4. lounge 5. entrance 6. wardrobe 7. guest room 8. dining room
9. kitchen 10. living room 11. terrace 12. apartment-1

一层 first floor

1 办公室 2 主卧 3 洗手间
4 卧室 5 公寓-1
1. office 2. master bedroom 3. restroom
4. bedroom 5. apartment-1

二层 second floor

E-E' 剖面图 section E-E'   F-F' 剖面图 section F-F'

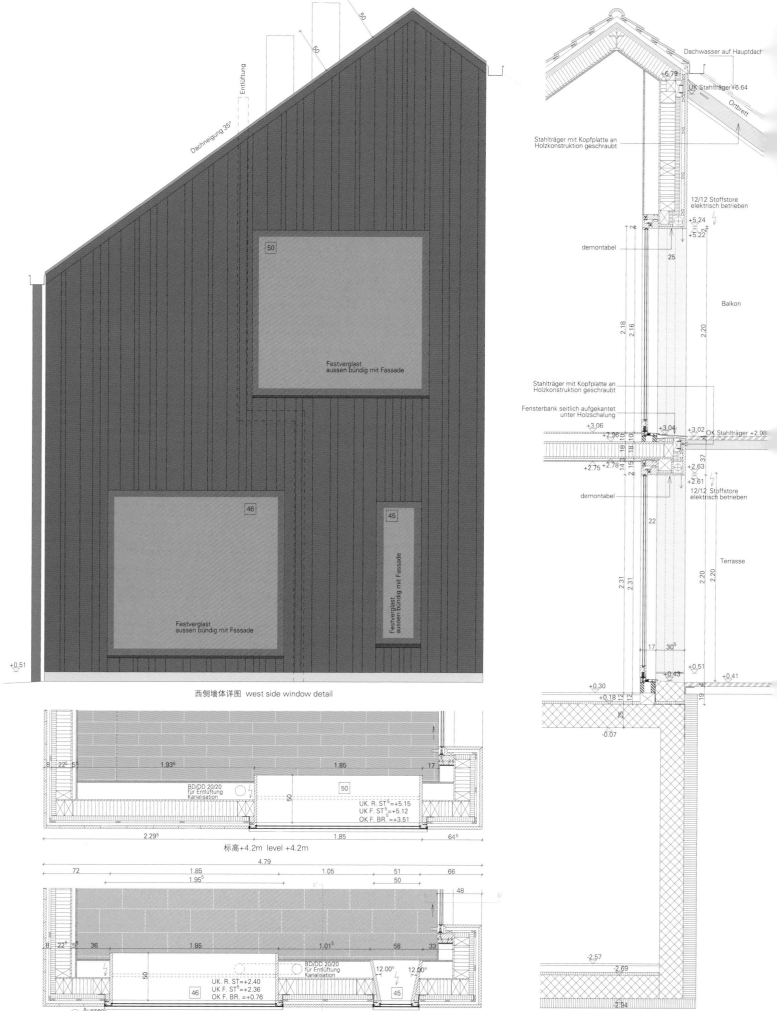

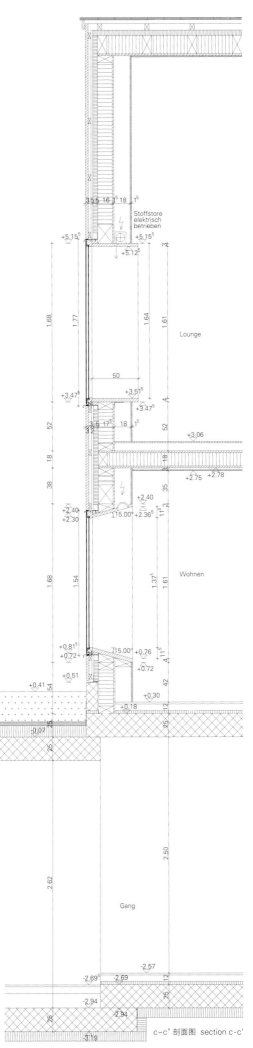

c-c' 剖面图 section c-c'

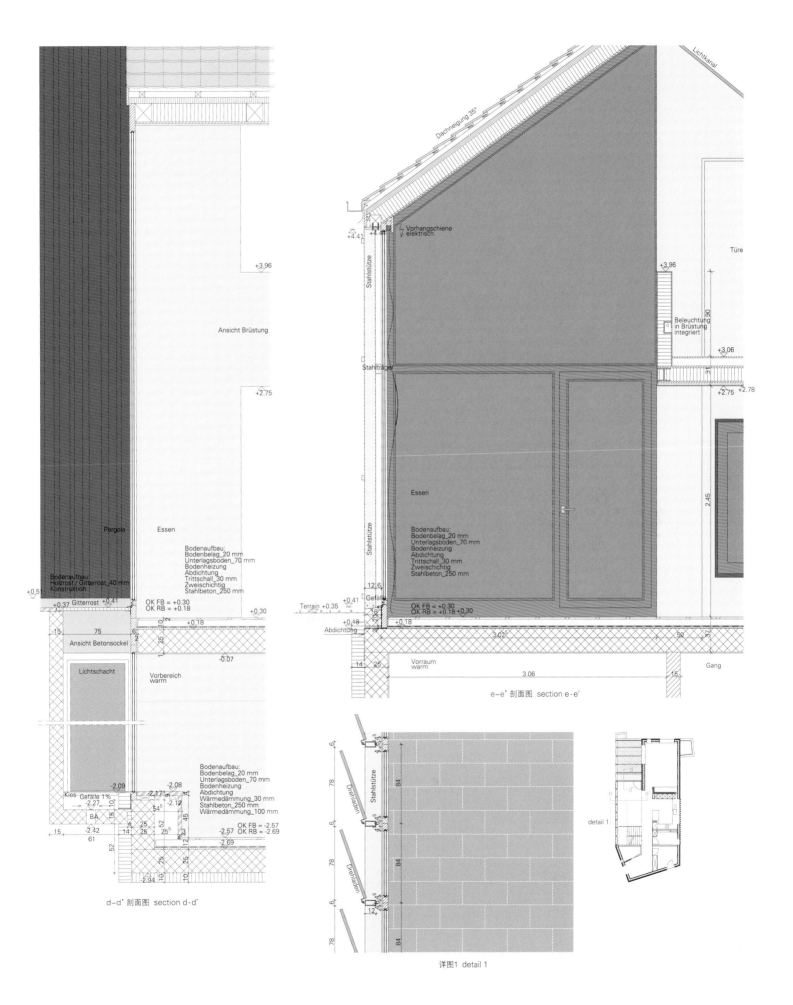

Torus

N Maeda Atelier

将天空转译在立面上

Torus的基本构思是一个双层结构,它是一个漂浮在低层、白色且半边不规则的盒状造型,由周边的玻璃和多孔铝镶板将其柔和地包裹在其中。

低层设计成透明开放式是为了实现其功能性需求的一个自然选择,其目的在于让潜在的客户和其他行人注意到美容院的存在,同时也是为了开辟出商店区域外侧的曲形多孔隔板围合的场地,以作为"遛狗"场地,使得小狗能够自由自在地在里面奔跑撒欢。

曲线形的墙体看上去线条较为自由,而实际上建筑师是基于细心地认知其所定义的"欢迎区域"来设计的。譬如说,小型区域需要向建筑场地外的城市背景开放。这些被切割的区域用作停车场、入口和室外设施区,这又造成了墙体的不规则的曲线形状,而这种形状正是这些施工所造成的。

由两层楼构成的上层单元沉重厚实的外表几乎像是一个重重武装的坦克,保卫着业主的居家生活,与下层的开放透明形成了鲜明的对比。近看,墙体的表面纹理像手工制作的陶器外观,而不是工业产品平整统一的质地。

表面近似陶的纹理是通过辛苦地在墙体上手工反复涂抹防水材料做成的。除了考虑到近距离观看的纹理,二层的外部设计同样融入了建筑师对于建筑远观效果的感受,它是通过将天空转译到墙体上而得以实现的。

这一转译过程如下:首先,在框架搭建完工的那天,建筑师在原地拍摄了一张场地上方天空的照片(阶段一);在那之后,经过提取加工,将其制成一张渐变的灰度图像(阶段二)。在小心翼翼地把它复制在墙面的4个墙边处后,这幅图像就被作为波形表面的等高线图来使用了;最后,建筑师用饰面灰泥将其进行覆盖。灰泥厚度(从0mm到30mm不等)的控制要取决于这些轮廓线——一个暗淡多云的天空质地就这样呈现在了人们面前。

完工后的墙体就像是138页下方照片里的景致一样,自然地呈现出一种天空中飘荡着一丝淡淡的云彩般的效果。

建筑师接着移到大型漂浮的盒子的部分,它似乎非常排斥周围的城市环境。

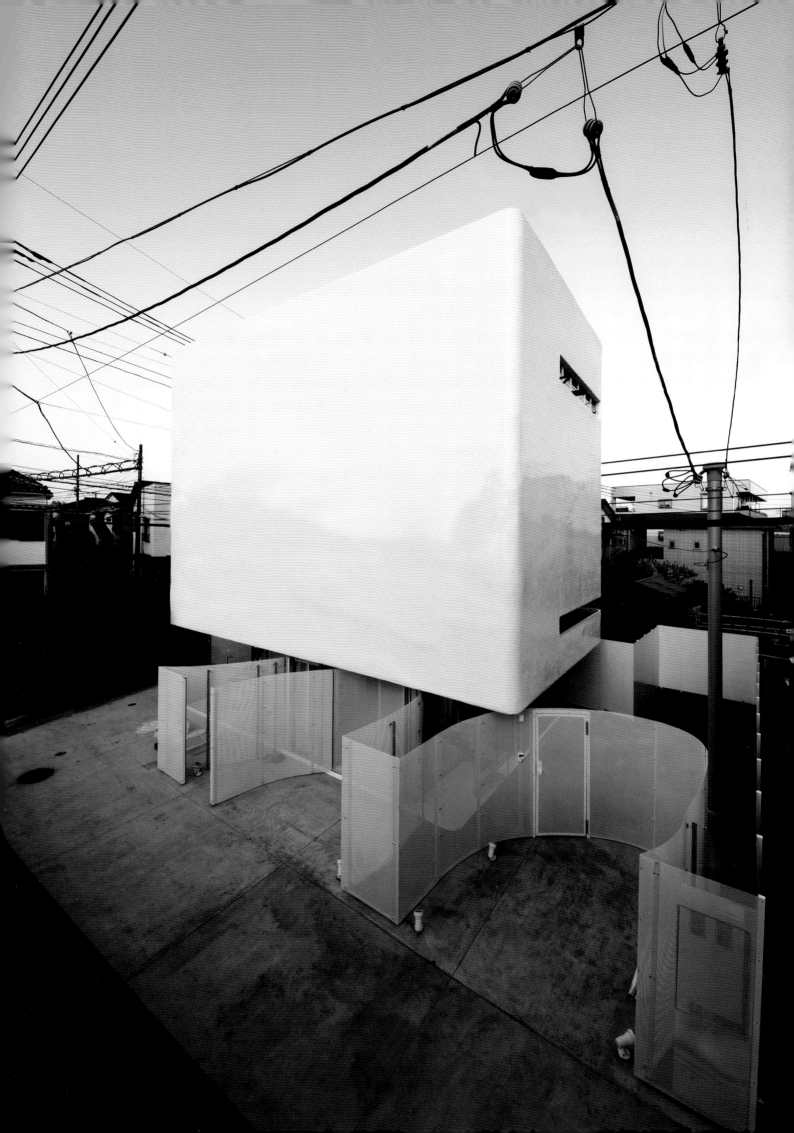

入口门旁边是一处小型空间：虽然它仍是一个室外的空间，但由于对比例和洞口进行的仔细处理，它在某种程度上拥有室内的气氛。

入口门内的正是大型盒子内的巨大体量。

很明显，形成这一"室外效果"的首要原因就在于屋顶悬挂着的那盏巨大的顶灯。然而这里其实还有另外一个不太为人所知的原因：凸凹不平的内墙饰面。

事实上，所用材料本身是在日本任何一家五金店都能买到的一种既普通又便宜的胶合板板条。为了让这种家居材料具有一种特殊的触感，建筑师将这些板条切割成每个宽度为200mm的细条，之后用手工把每条上面较软的部分进行剔除，从而使其表面上较硬的纹理能够被凸显出来。随着最后白漆的上色（该过程需要采取一项特殊的准备措施来确保吸水和不吸水表面各部分都能够均匀上色），先前毫无特色的材质已然被改造成如今独一无二的饰面材料。

Torus

Transcribing the Sky onto the Facade

Basic composition of Torus is a bilayer structure consisted of a white, half-amorphous box floating on the lower layer softly surrounded by glass and perforated aluminum panels.

The transparency and the openness of this layer is a natural solution for the functional requirement to expose the presence of the salon to prospective customers and other passers-by, as well as to open up the ground outside the shop area surrounded by the curved perforated partitions as "dog-run" field where dogs can freely run around.

The apparently free line of the curved wall is actually based on the architects' careful recognition of what they call the "welcoming zones", i.e., the pocket areas required to open up to the urban context outside the building site. The cutting-outs of such zones like parking, entrance and spaces for outdoor equipment have resulted in the irregular curve of the wall as the output of such operations.

The upper unit containing two floors within presents a sharp contrast to the open, transparent lower layer with its weighty, massive appearance almost like a heavily-armed tank defending the rather indoorsy life of the client family. With a closer look, the surface of the wall shows a texture similar to a handmade pottery instead of that of a flat, uniform industrial product.

The pottery-like texture is the result of the painstaking manual operation of repeatedly spreading the waterproof material onto the wall. Beside such a consideration to the close-up texture, the exterior of this second layer also involves the architects sensitivity to the longer-distance outlook of the building, which is realized by an operation of transcribing the sky onto the wall.

The transcription process is as following: first, the architects took a picture of the sky right above the site on the day of framework completion (phase 1); the picture was then abstracted into a grayscale gradation graphic (phase 2), which they applied as the contour map of the undulating surface by carefully duplicating it on the four sides of the wall; finally, they covered the surface with finishing mortar while controlling its thickness (varied from 0mm to 30mm) based on the contour lines – and thus the ambiguous cloudy sky texture emerged.

The finished wall naturally takes on a feature of the sky with wispy clouds, as shown in the pictures below.

The architects move up into the massive floating box, which appears extremely exclusive of the surrounding urban context. Beside the entrance door is a small pocket space: although it is still an open-air space, it somehow bears an indoor atmosphere due

阶段一 phase 1

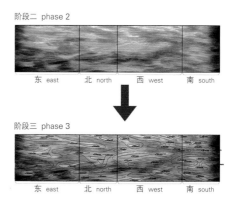

阶段二 phase 2

阶段三 phase 3

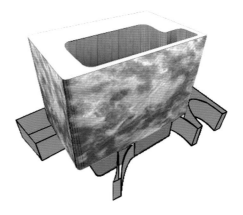

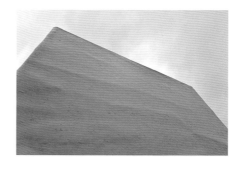

项目名称：Torus
地点：Saitama, Japan
建筑师：N Maeda Atelier
首席建筑师：Norisada Maeda
结构工程师：Ryozo Umezawa 承包商：Iwamotogumi
用途：private residence, pet-shop
用地面积：169.14m² 总建筑面积：57.42m² 有效楼层面积：145.32m²
材料：plywood, waterproofed paint 竣工时间：2013
摄影师：Courtesy of the architect - p.136, p.137, p.139[bottom], p.141, p.142, p.143, p.145[bottom]
©Studio Dio (courtesy of the architect) - p.139[top], p.144, p.145[top]

to the careful treatment of proportions and openings.
Right inside the entrance door is a huge void within the massive box.

The prior factor of this "outside effect" is obviously the gigantic top light on the roof, but there is another, rather obscure one: the rugged interior wall finish.

The material itself is actually an ordinary, cheap plywood panel available in any hardware store in Japan. To give the particular tactile quality to this daily material, the architects cut the panels into narrow boards of 200mm width each and then manually removed the soft parts from each and every board to let the hard grains stand out on the surface. With the finishing white paint (which needed a special preparation to evenly paint over the water-absorbing and non-absorbing portions of the surface), the ordinary material has been turned into a unique finishing material like this.

N Maeda Atelier

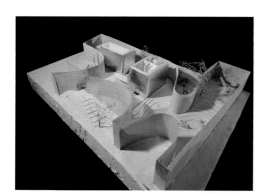

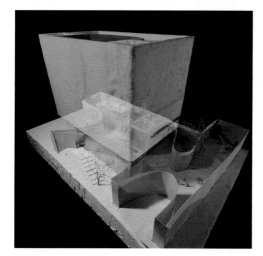

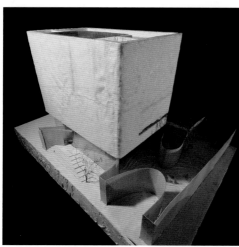

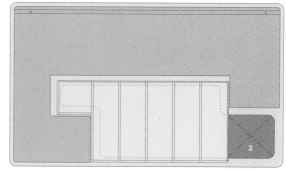

1 顶部采光区 2 上空空间
1. top light 2. void
屋顶 roof

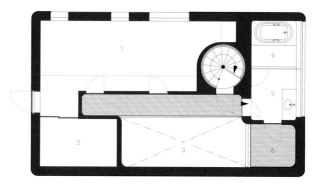

1 卧室 2 卫生间 3 上空空间 4 浴室 5 化妆室 6 露台
1. bedroom 2. WC 3. void 4. bathroom 5. powder room 6. terrace
三层 third floor

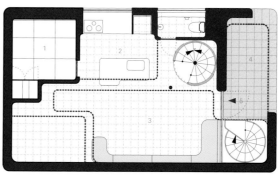

1 日式房间 2 厨房 3 休息室 4 露台 5 入口
1. Japanese room 2. kitchen 3. lounge 4. terrace 5. entrance
二层 second floor

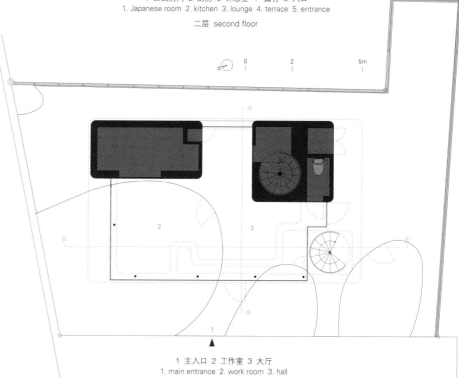

1 主入口 2 工作室 3 大厅
1. main entrance 2. work room 3. hall
一层 first floor

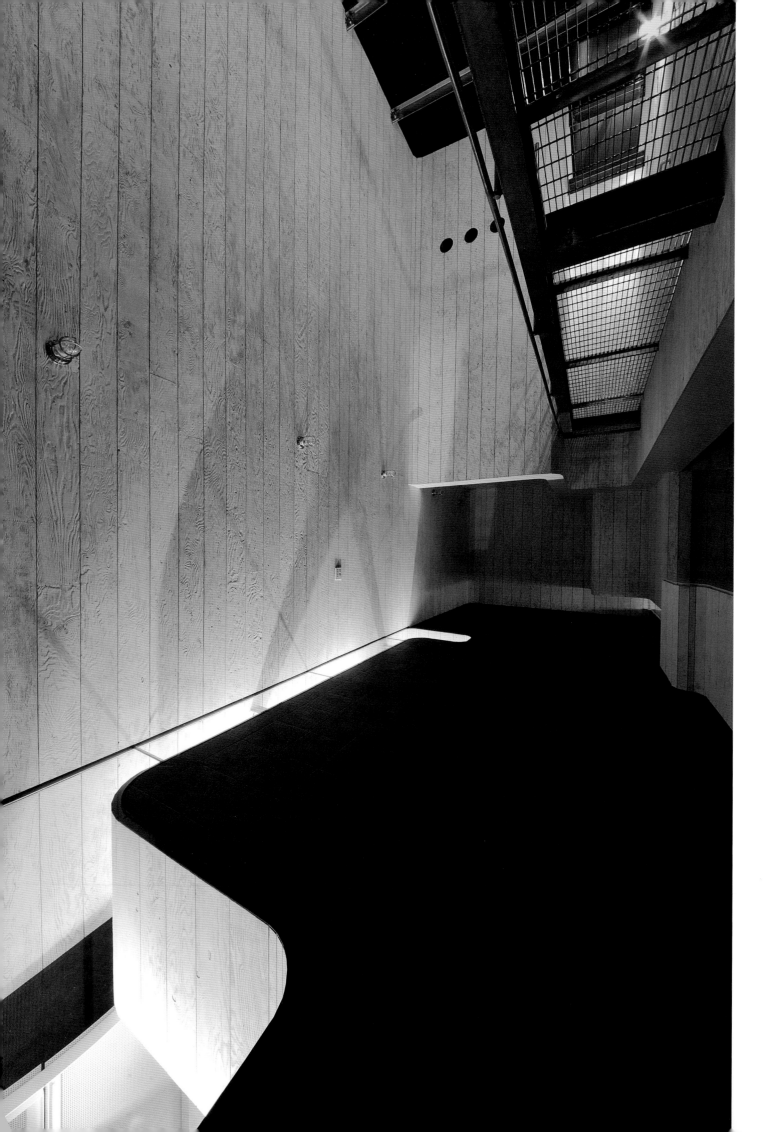

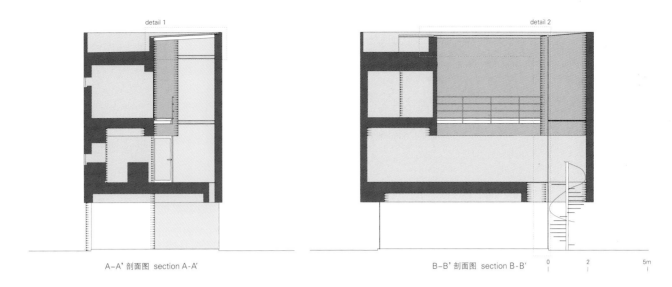

A-A' 剖面图 section A-A'   B-B' 剖面图 section B-B'

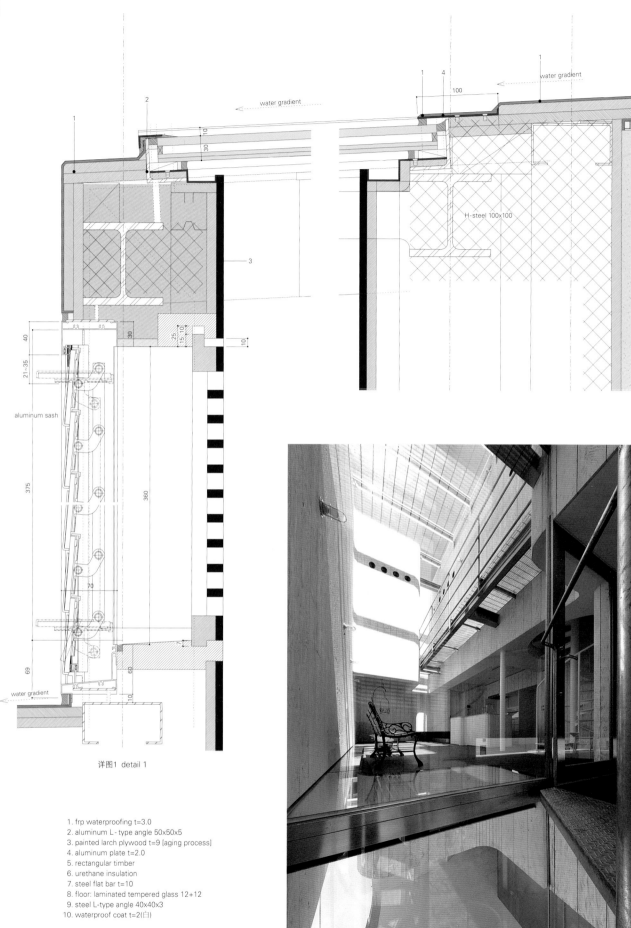

详图1 detail 1

1. frp waterproofing t=3.0
2. aluminum L - type angle 50x50x5
3. painted larch plywood t=9 [aging process]
4. aluminum plate t=2.0
5. rectangular timber
6. urethane insulation
7. steel flat bar t=10
8. floor: laminated tempered glass 12+12
9. steel L-type angle 40x40x3
10. waterproof coat t=2(白)

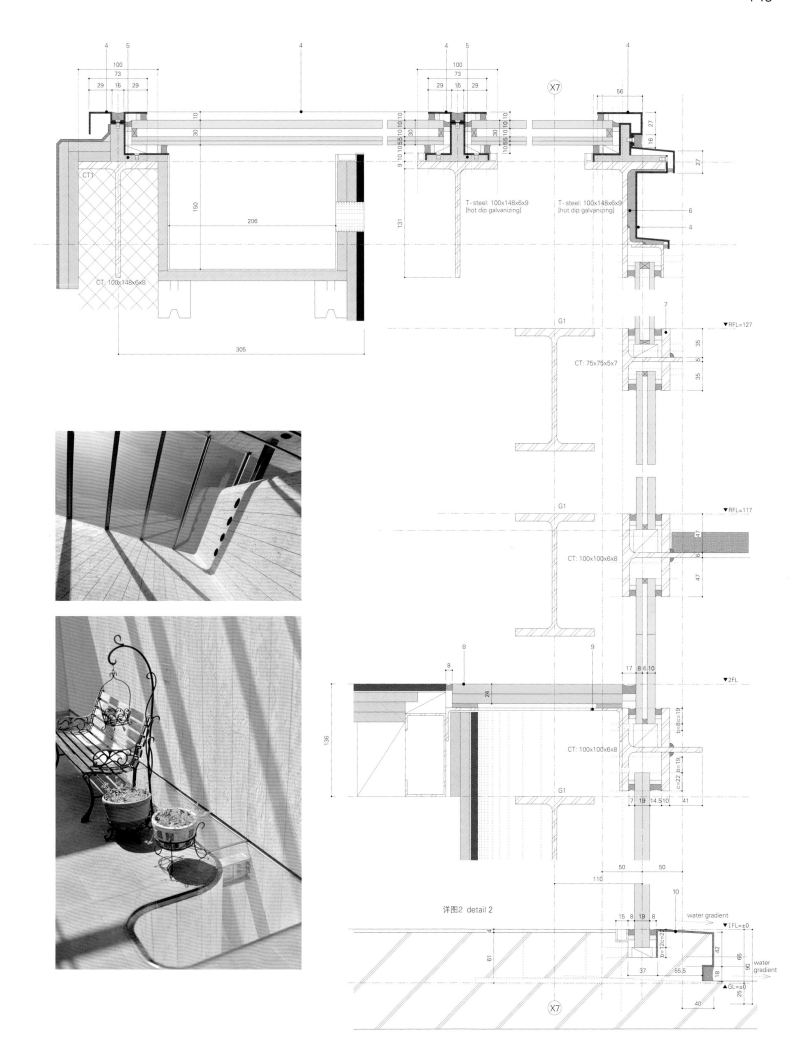

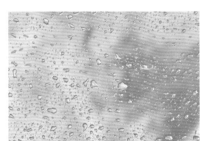

莫托萨布公寓sYms

Kiyonobu Nakagame Architects and Associates

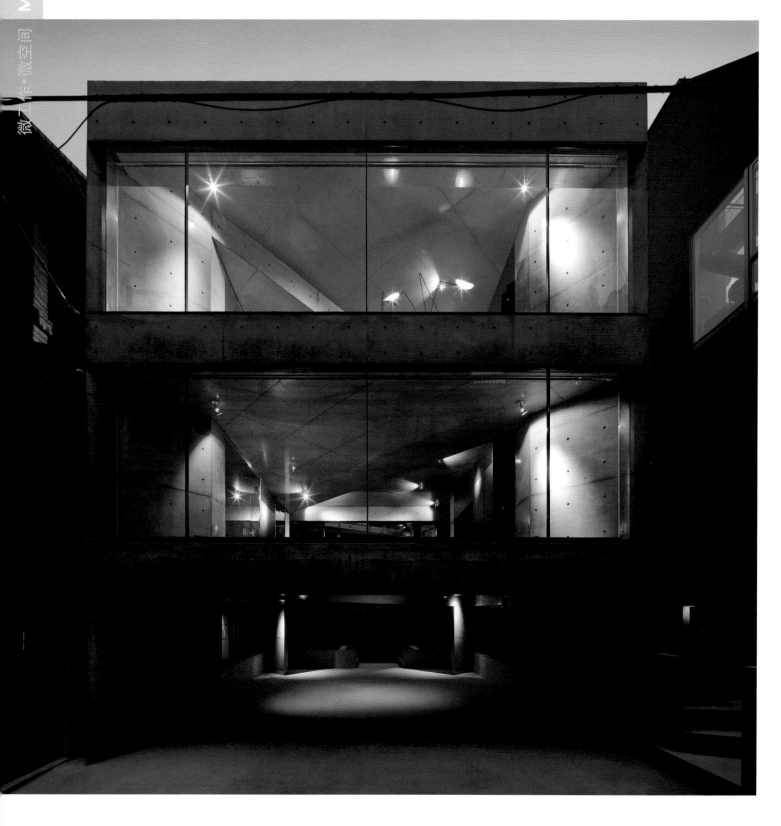

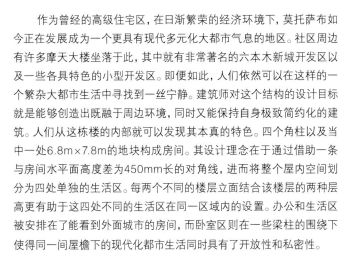

作为曾经的高级住宅区,在日渐繁荣的经济环境下,莫托萨布如今正在发展成为一个更具有现代多元化大都市气息的地区。社区周边有许多摩天大楼坐落于此,其中就有非常著名的六本木新城开发区以及一些各具特色的小型开发区。即便如此,人们依然可以在这样的一个繁杂大都市生活中寻找到一丝宁静。建筑师对这个结构的设计目标就是能够创造出既融于周边环境,同时又能保持自身极致简约化的建筑。人们从这栋楼的内部就可以发现其本真的特色。四个角柱以及当中一处6.8m×7.8m的地块构成房间。其设计理念在于通过借助一条与房间水平面高度差为450mm长的对角线,进而将整个屋内空间划分为四处单独的生活区。每两个不同的楼层立面结合该楼层的两种层高更有助于这四处不同的生活区在同一区域内的设置。办公和生活区被安排在了能看到外面城市的房间,而卧室区则在一些梁柱的围绕下使得同一间屋檐下的现代化都市生活同时具有了开放性和私密性。

Motoazabu Apartment sYms

Once known as a high-class residential area, Motoazabu is developing into more of a modern diversified metropolitan region resultant of the economic boom. The neighborhood consists of skyscrapers like the famous Roppongi Hills development alongside smaller unique small developments. But within the chaotic big city life of this area people can still find tranquil quietness. What the architects aimed to do with this structure was to create something that would blend with its surroundings and maintain absolute simplicity. The true character of the building can be found on the inside. A structure consisting of 4 corner columns with a footprint of 6.8m x 7.8m comprises one room. The design concept takes in a diagonal line of 450mm in room level difference which lends itself to create 4 distinct living areas all within a single space. The 2 different elevations of the floor combined with 2 different ceiling heights lend the space to create 4 distinct living areas all within a single space. The office and living areas take up the stage with views of the city and the bedroom space surrounded by post beams provides openness and a sense of privacy to modern city life in one room.

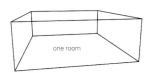

四种功能
(起居室、餐厅、卧室和办公室)
4 functions
(living, dining, bedroom, office)

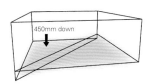

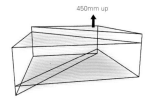

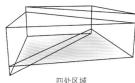

四处区域
(起居室、餐厅、卧室和办公室)
4 areas
(living, dining, bedroom, office)

项目名称:Motoazabu Apartment sYms
地点:Tokyo, Japan
建筑师:Kiyonobu Nakagame Architects and Associates
用地面积:100m²
总建筑面积:59.56m²
有效楼层面积:139.49m²
结构:reinforced concrete
设计时间:2011.1—9
施工时间:2011.10—2012.6
摄影师:©Shigeo Ogawa (courtesy of the architect)

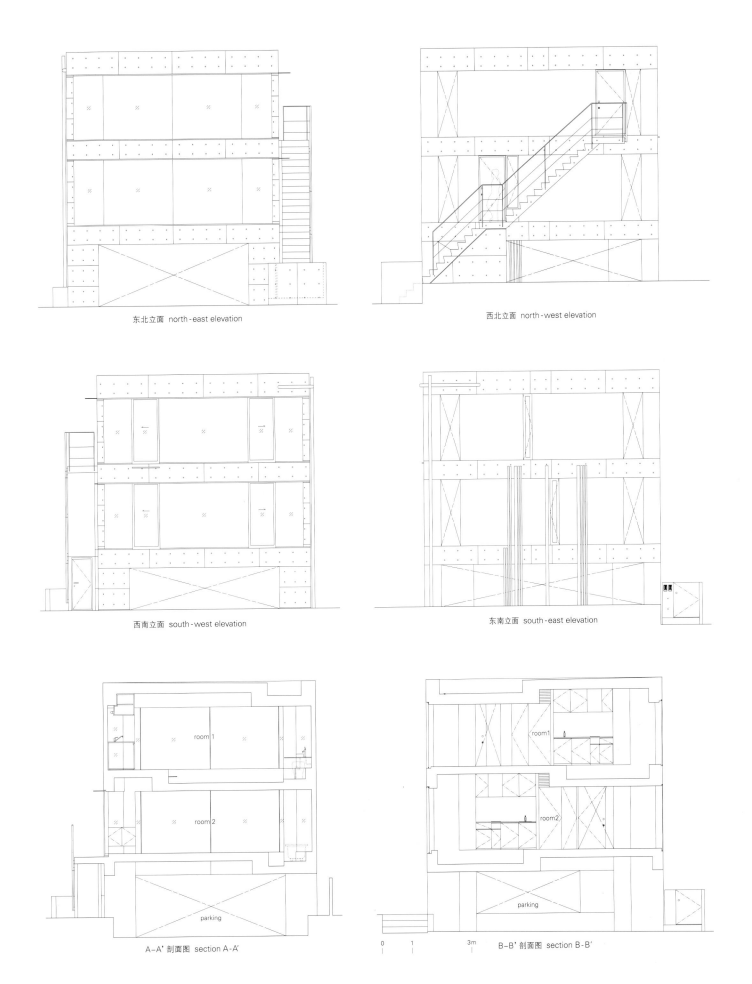

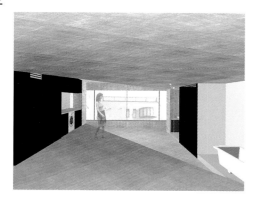
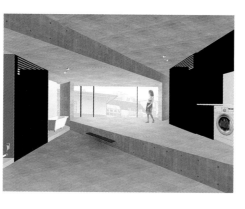

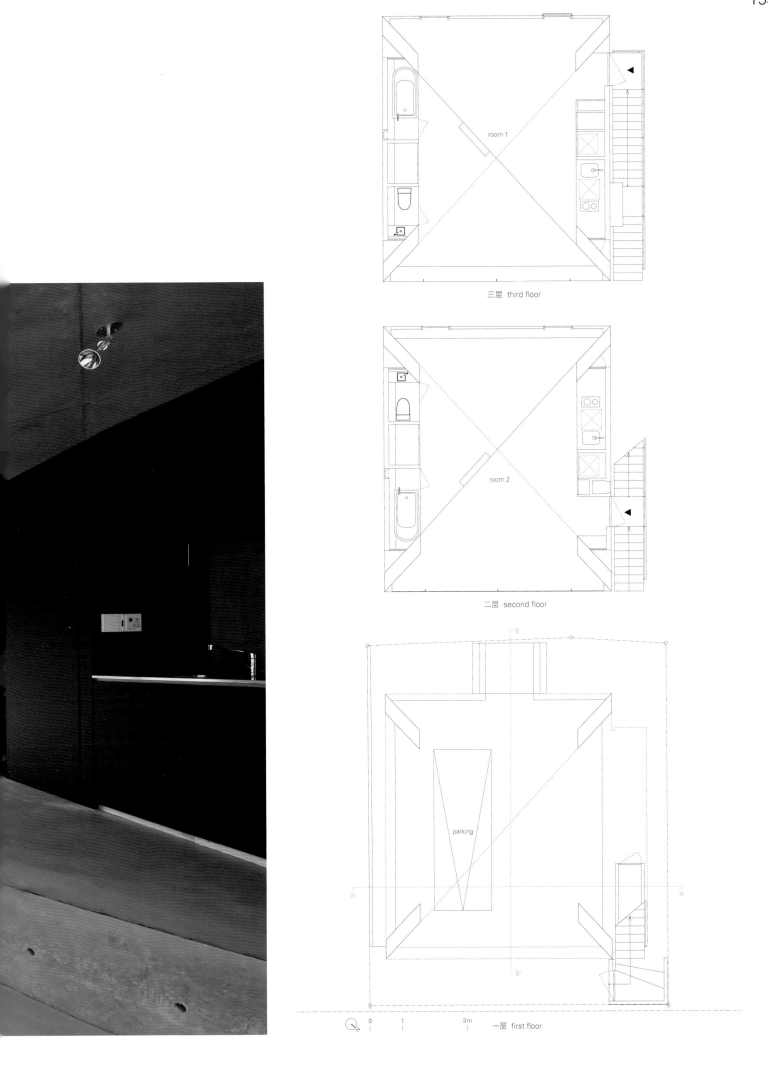

三层 third floor

二层 second floor

一层 first floor

良性发展的城市建筑
Virtuous Urban Pieces

如今，当人们谈论起城市发展时，往往会不可避免地涉及到一些与私有的公共空间相关的主题。从这个意义上来说，我们会发现有许多诸如此类的例子。其中，有从雷姆•库哈斯所描述的"普通城市"的影像到一些更加繁复的城市集聚性设计，如设计师史蒂芬•霍尔所设计的来福士广场以及Rogers Stirk Harbour+Partners建筑事务所设计的NEO Bankside豪华公寓。在之前的两个例子当中，建筑师的主要挑战就是去找寻他们所设计的建筑物在整个城市中的正确定位，与此同时，还要以一种即便在最小规模中都能够凸显出其自身意义的方式来清晰地展示出这一项目的构成元素。为了达到这一目标，塔楼以及典型的隔离式结构的使用均能够体现出这些建筑的独特风格。对于多规模或者是多系统型的建筑以及高层来说，便利性和微都市化都必须被用作附加性的关键词，从而对相关类型的设计作品进行解读。

这些特点不仅确保了这些作品本身达到一定的标准，同时也保证了人们在这样不同寻常而又实效的当代城市中的住房以及生活。此外，他们也将自身看作是一个正在展望未来的城市中的一个战略性要点。

Nowadays talking about urban development inevitably involves themes related to privately owned public spaces. In this definition can be grouped several examples, from simulacra in the "Generic City" described by Rem Koolhaas to more sophisticated urban assembles like the Sliced Porosity Block by Steven Holl Architects and the NEO Bankside by Rogers Stirk Harbour+Partners. In the last two cases, the architects' main challenge was to find the right role for their buildings into the city and, at the same time, articulate the project elements in a way that they can be significant at the smallest scale. Achieving such a goal through the use of towers, quintessential isolating typology, shows the exceptionality of these architectures. To multi-scale/system architecture and high rise typology, accessibility and micro-urbanism must be added as complementary keywords to interpret these type of works.

These characteristics qualify not only these examples but also the housing and living in the contemporary conurbations when they are remarkable and effective, and set themselves as strategic points for a city that looks toward the future.

When it comes to capture and express the contemporary vibe of an urbanism-related phenomenon, the challenge faced by the observer is to discern common themes present in architecture and urban fabrics that are very far from each other. For example, taking into consideration the Sliced Porosity Block project by Steven Holl and the NEO Bankside project by Rogers Stirk Harbour+Partners, differences on visual impact, strategy on the site, program, aspiration, and scale may be evident. What is less obvious is how these two projects share the same attitude towards the urban composition, the keystone of both developments. Through the relationship between the volumetric components, the landscape and the rest of the city these projects acquire meaning and quality.

First of all it must be noticed that despite the use of the tower type in both projects, the results have a strong positive impact on the urban composition. In fact, not only do they generate an interesting ensemble in which the volumes are not isolated, but also they engage the external ground level as if they were part of a medium density development. As a consequence of generating an elaborate complex of towers, both schemes gained a role in the city shape that is less domineering and more open to the future growth of the conurbation. After the destruction of the urban block by modernism, which replaced a system of spaces with a system of bodies, as stated by Leon Krier in his *the Reconstruction of the City*, these examples seem to observe two stages of the same inverse phenomenon. In both cases the design process is formed by two phases: the definition of a clear precinct through a single piece of architecture and the subsequent slicing of smaller volumes, with the intention of revealing the "inside" of the urban block or creating visual links with the surroundings. Following this strategy, the wider urban realm works in a more cohesive way and the new, architecturally cutting edge developments act as high quality parts in it.

Obviously the role of the two developments appear to be completely different analysis of the surroundings. Rogers Stirk Harbour+Partners' project is located in one of the most significant areas of London, adjacent to the Tate Modern and with views towards St. Paul's Cathedral, hence it is almost natural that it acquires some traits of a linking area, rather than being a place of

NEO Bankside豪华公寓/Rogers Stirk Harbour + Partners
来福士广场/Steven Holl Architects

良性发展的城市建筑/Simone Corda

　　当涉及到对城市化相关现象同期感受的获取和表述方面时，观察者所面临的挑战就是要辨别彼此之间距离较为遥远的建筑以及城市肌理中那些相同的主题。例如，就拿史蒂芬·霍尔所设计的来福士广场以及Rogers Stirk Harbour+Partners建筑事务所设计的NEO Bankside豪华公寓来说，二者在视觉冲击、选址策略、设计安排、期望值以及规模方面都存在有较为明显的差异。略为不太显著的一点是，这两个工程在面对着城市化构成这一要素时，究竟如何做才能使双方都能持有相同的态度。通过其体量构成、景观及城市其他部分之间的关系，这些项目由此体现出了自身的意义以及品质。

　　首先必须注意到的一点是，尽管两个建筑中都有塔式设计的运用，其最终结果都对城市化构成产生了强大的、积极的影响。事实上，它们不仅生成了一个有趣的整体效果（体量没有隔离开来），同时也参照了外部地表水平，使其好像是中等密度开发区的一部分。创作出这些精美而又复杂的塔式结构的结果是，这两项设计在并不张扬的同时也在未来都市化进程越发开放的城市中都赢得了自己的一席之地。城市化用一系列实体取代了一系列的空间，从而对城市街区造成了一定的破坏，就像Leon Krier在他的《城市的重建》一书中所提到的那样，从这些实例当中我们可以发现在同一个逆向的现象中所存在着

NEO Bankside/Rogers Stirk Harbour + Partners
Sliced Porosity Block/Steven Holl Architects

Virtuous Urban Pieces/Simone Corda

的两个不同的阶段。这两项工程的设计过程历经了两个不同的时期：通过一个简单的建筑结构以及其后对较小体量所进行的切割来给相应明确的区域下一个定义。这么做的目的是为了展示城市街区的"内部"，或者是创造出一种与周边环境相关的视觉联系。鉴于该项策略，更加宽泛的城市圈以一种结合得更为紧密的方式得以运行，而这一建筑上新型的尖端开发区则成为其中的高品质组成部分。

　　显而易见，这两个开发区的角色似乎都是对其各自周边环境的完全不同的解读。Rogers Stirk Harbour+Partners建筑事务所设计的项目坐落于伦敦最著名的地区之一，旁边就是泰特现代美术馆，与圣保罗大教堂相望。因此，相较于成为某个目的地来说，这里非常自然地形成了一个连接区的特色。虽然这些外部路径并不能够被当作是主要的功能性通道，但这一过渡性的功能依然通过提高该街区渗透性的路径而被清晰地表达了出来。最重要的路段依然是夏街，那里的塔楼群和马路对面的建筑物为美术馆形成了一个可视的大门。当赫尔佐格&德梅隆建筑事务所设计的泰特新派系项目即将完工之时，这一大门将被改造成一个能够部分覆盖住周边建筑物的有趣屏障，就这样，之前的视角也会有所改变。这一地区的历史与文化遗产相关性同样也可从建筑的高度上体现出来，这一高度的上下起伏一般也要取决于周边其

arrival. The transitional aspect is expressed quite clearly by the external paths that improve the permeability of the block, even though they cannot be considered the main functional passageway for the general public. The most important road remains Summer Street, where the towers create a visual gate for the Tate with the building on the opposite side of the road. This perception is going to change when the Tate's new wing by Herzog & de Meuron will be completed, transforming this gate into a teasing screen partially covering the new building. The historical and cultural heritage relevance of the area can be also traced in the heights of the construction, which step up and down according to the surrounding buildings. The architects lowered the heights toward the Hopton Gardens, while they reach the maximum level in correspondence of Summer Street.

Steven Holl had to face a completely different challenge in their Sliced Porosity Block project, located in Chengdu, China. As he affirms in the project's promotional video, the Chinese culture "understands the urgency of building for the future", hence its design is based on a projection of the city yet to be. The urban block stands out from the surroundings, with a scale accorded to the newer skyscrapers rather than the medium density volumes that characterize the immediate context. The project also contradicts its anonymous and repetitive lining up of the buildings on the East-West axis, proposing a more articulated composition around a plaza. In order to light the vast plaza, the buildings are placed around the boundaries, open to the South and sliced to allow the light to cut through. An elaboration of the conceptual reference of the Rockefeller Center was Holl's main aim, specifically the design of a complex with no iconic buildings, able to shape the focal public space; this also happens in the NEO Bankside project.

Therefore two approaches to the space have been developed by the two architectural teams. NEO Bankside proposes the repetition of a regular six-sided apartment block with different heights. To a viewer the facades of each building appear in succession to each other with different orientations, colors and textures, opening visual corridors in a highly dynamic space. The tower exposed structures, characterized by their diagonals braces, not only allow for highly flexible interior configurations, but also increase the

照片提供：© Hayes Davidson and Herzog & de Meuron

赫尔佐格&德梅隆建筑事务所设计的新派系作品泰特现代美术馆坐落在与NEO Bankside豪华公寓相邻的地区
the Tate's new wing by Herzog & de Meuron, being built adjacent to NEO Bankside

他的建筑物群。建筑师拉低了朝向霍普顿花园的高度，而这已经达到了与夏街相对应的最高高度。

史蒂芬·霍尔早前在中国成都的来福士广场项目中不得不面对一个极度困难的挑战。就像他在项目的推广视频中所断言的那样，中国的文化"理解未来建设的紧迫性"。由此，这个设计是建立在对这座城市未来预测的样貌的基础之上的。比起当前环境下较具特色的中等密度的楼群来说，这一城市街区在周边环境的映衬之下有着更符合新建的摩天大楼的规格。该项目也会与自身毫无特色且重复的罗列在东-西轴线上的一些建筑物产生矛盾，因此提出了建筑群要围绕着某个广场的这样一个连接性设计。为了凸显这一大型广场，建筑群被放置在了面向南方的边界地带，并且加以分割，使阳光可以自然照射进入其中。对洛克菲勒中心的概念性参考进行一个详尽的规划是霍尔的首要目标，特别是要对一个没有标志性建筑的综合体，从而使它能够成为公众区域的中心进行设计；这一点对于NEO Bankside项目来说也是同样的。

就这样，两个设计团队采取了两种方案来对这一区域进行设计。NEO Bankside项目被规划为不同层高但完全相同的规则六边形公寓楼。对于一个观众来说，每栋建筑物的立面看上去都处在一个有着不同朝向、色彩以及质地的序列之中，由此打开了一个位于高度动感空间内的通道。塔楼结构的特色是运用了对角撑，它不仅允许设计具有高度弹性的内部布局，同时也增加了一定的动感。塔楼之间的紧凑性以及它们整齐的行列积累了丰富的视觉体验，使其构成一个整体。从另一方面来说，来福士广场项目是一个经过了叠加和切割处理的一体式单独结构。它是在支持建筑物顶部设置标志性构件的传统设计基础上所进行的一项创新发展。在这个项目中，常规性和重复性被一种更加全盘性的方式所替代，它将所有的元素一并放置在了明显离雕塑更近的位置上。建筑物带有倾斜度的表面可以对一些空间进行打开和关闭，因此，拥有了这些空间，楼盘的分离进而使得该布局具有了高度的动感。

然而，这些项目中也存在有一些相似之处：尽管建筑师们都对体量构成给予了极大的关注，但这些建筑物各自的外表却为自身带来了一定的附加值。由此，两位建筑师再度采用了截然不同的方法。NEO Bankside项目的塔楼有一个多层的表皮，而结构置于木材复合板、百叶窗以及玻璃嵌板组成的体系之上，从而营造出一种历经提炼的、光滑的亮度。

这一组合完美地将建筑群和该地区19和20世纪的工业遗产融合在了一起。与构造上的清晰度相反的是，来福士广场项目体现的是一种平坦的外表皮，其中的洞口网格形成了一种抽象的多变模式。抗震

sense of movement. The mutual proximity of the towers and their alignments organize a visually rich experience into a unity. On the other hand the Sliced Porosity Block is formed by a coherent unique volume, folded and sliced, which is an innovative development from the traditional configuration of a base that supports iconic objects on top. In this case the regularity and the repetition are replaced by a more holistic way of putting the elements together, definitely closer to sculpture. Therefore the separation of the blocks achieves a highly dramatic configuration with the space that gets opened and closed by the building's slanted surfaces. Nevertheless there are similarities between the projects: even though the architects gave great importance to the volumetric composition, the skins of the buildings provide the added value. Once again the two architects followed diverging approaches, the NEO Bankside's towers have a multilayered skin, where the structure is superimposed on a system of timber clad panels, window louvers and glass panels, creating a refined, slick lightness. This combination perfectly merges the buildings with the 19th and 20th century industrial heritage present in the area. Opposed to this constructive clearness, the Sliced Porosity Block shows a flat skin in which the grid of the openings creates an abstract variable pattern. The orthogonal scheme is shot through by the earthquake-resistant diagonal braces so that this additional level of complexity reduces the obsessive character of the grid. The slanted short sides of the buildings have an impressive strength thanks to their glass elevations that look as though they were carved from a single original volume showing their real core.
In both projects the non-orthogonal surfaces underline and visually accelerate the sense of passage from the public space inside the development to the rest of the city. This porosity expresses the need for the project to become part of a bigger system, nevertheless the definition of precincts and their recognizability are key points. In reality, although the two design teams have shaped the external space in a different way, according to the role and to the dimension of it, they both referred to the concept of a micro environment partially separated from the rest. In London the landscape designers Gillespies create a green oasis in the middle of the city. Linear gardens separate the public paths from the pri-

纽约的祖科蒂公园是自20世纪后半叶以来存在至今的一处非常著名的公共空间

Zuccotti Park, a remarkable public space in New York that has been going on since the second half of the previous century

对角撑形成了一个相互垂直的结构,以至于这个综合体的附加层可以降低整个网格所带来的压迫性。建筑物较短并且倾斜着的这一面由于设有透明的玻璃电梯而显得具有一种令人印象深刻的力量。这些电梯看起来就像是从显现出其真实核心的原始建筑中所切割出来的一样。

在这两个项目中,对公共空间内通道的识别力由于不垂直的表面而有所加强,并且从视觉上来说也有所加快。这一公共空间位于这座城市中其他片区的开发区以内。尽管这些区域的定义和可辨识性是关键,然而其空隙率依然表达出这项工程对于成为更大一个系统中一部分的一种需要。在现实中,虽然这两个设计团队用各自不同的方式对外部空间进行了规划,然而根据其定位以及自身面积,二者均提到一个与其他环境相隔离的迷你环境的概念。在伦敦,Gillespies景观设计事务所在城市的中心创建了一片绿洲。线性的花园将公共通道从一个私人美化的区域分离出来,从而形成了一处柔和的过渡区域。就这样,不同用法可以同时被投入使用:当居民可以享受这样一个安全祥和的空间时,参观者也得到了一个靠近这里的机会,同时也可以在这样一个半隐蔽的美妙世界中尽情畅游。从相反角度来说,对于各种元素的处理和植入过程同样有着一个既复杂又统一的功能。事实上,从一个基本的物质角度来说,这个公园建立起了与旁边阿伯顿公园之间的连接。在分析其细节的语言和规模时,它在这个小范围中营造出一种和谐。

成都的这一公共区域由几种类型的地点构成:楼内构成元素,与三处水的特点有关的三个广场以及史蒂芬·霍尔口中的"悬浮性注入功能"元素,三个与该城市最伟大的诗人杜甫以及高科技和地方历史有关的凉亭。与伦敦的场地不同,该项目的范围允许对建立在不同水平面的外界进行一个更加复杂的连接。主要的区域是一个三层的中央广场,由杜甫诗中所描述的三处峡谷来进行诗意的连接,以构成其结构。公共区域的其他部分也设有亭子,一些雕塑元素位于主体量的上层,作为建筑中的建筑而存在着。

因此,这两项工程引发了一些特有的趋势:对于可融合于周边城市的独特建筑的需求以及将一个公共区建成一个私人空间的改造。

这两项工程是自20世纪后半叶以来非常引人瞩目的杰出实例,这种理念形成了一种得以持续的工艺流程。纽约的祖科蒂公园就是其中的一个产物。它们之所以能够打动人心,是因为它们与已建成的城市有着亲近的关系。尽管其中一个项目只是战略性地为其修建了一处区域,但其他的都被嵌入在其他系统压缩在一起的一个地区当中。

vate landscaped precincts forming a soft threshold. In this way different uses are allowed at the same time: while the residents can enjoy a secure peaceful space, the visitors have access and can cross this semi-hidden pleasant world. From an opposite perspective, the treatment of the elements and the planting also has a complex unifying function. In fact, from a basic physical point of view, the garden establishes a continuity with the adjacent Upton Gardens, while, analyzing the language and the scale of the details, it creates a harmony based on the small dimension.

The public realm in Chengdu is formed by several types of places: the components inside the buildings, the three plazas associated to the three water features and the elements that Steven Holl called the "suspended injected functions", three pavilions related to the city's greatest poet Du Fu, the High Tech and the provincial history. Unlike the London site, the dimension of the project area allows a more complex articulation of the external space, based on the change of levels. The main area is a three level central plaza, structured to hint a poetical connection between the "Three Valleys" described in a Du Fu's poem. Other parts of the public space are the pavilions, sculptural objects located on a superior floor of the main volumes as buildings inside buildings.

Hence interesting trends arise from the two projects: the need for exceptional architectures that are able to merge with the city and the creation of a public realm into a private space.

The two projects are outstanding examples of a process that have been going on since the second half of the previous century, and that generated other remarkable places like Zuccotti Park in New York. They are striking because of their intimate relationship with the established city, even though one of them builds a strategic place for it, while the other is inserted in an area that is almost compressed by other systems. Simone Corda

NEO Bankside大楼

Rogers Stirk Harbour + Partners

该项目位于伦敦泰晤士河边区域（Bankside）的中心地带。在那里，许多文化、商业和住宅区以及一个繁荣多元化的环境共同存在于一个形成于中世纪的街道网内。场地离泰晤士河很近，正对着泰特现代美术馆的西侧入口。钢架玻璃小亭子完美地与河岸景观融合在一起，从而也映射出这里作为工业区的历史过往——同时也引领人们进入这一令人兴奋的住宅区。NEO Bankside大楼由所有建筑内12层到24层的217个住宅单元构成。该项目中所有楼房的设计线索均来自直接的外部环境，它代表了整体品质——而不是个别的组成部分，以产生戏剧性的效果。

风景优美的小树林清晰地划定出两条可以穿过这一区域的公共路线。该区域将现有的景观从泰特美术馆外的河边公园一直延伸到了南华街。

NEO Bankside项目周边独创性地设置了四个六角凉亭，其目的在于提供给居民更宽阔的居住空间，并且让他们享受到最充足的日光浴。这些雅致祥和、美景环绕的花园完美地与泰特美术馆和周边环境融合在了一起，使得人们可以每天来这里游玩。同时，居民们也有了一处安全且私人的空间供自己享用。从一间东北朝向且有三个卧房的居室望去，人们会惊喜地发现包括圣保罗教堂、伦敦城、泰晤士河、泰特美术馆以及千禧桥在内的众多美景。而从一个西南朝向的房间望去，人们便可以看到伦敦眼、大本钟以及英国的国会大厦。首都主要的商业、教育、购物以及娱乐中心全部在视线所及的范围之内。

总体的设计映射出该地区早前在19世纪和20世纪的工业化元素，并以一种现代的语言来对其做出回应。它将当地建筑特色中的着色及其选材都做了很好的解读。更加明确的是，这一提议试图获取一种现代的建筑化语言从而能够对当地环境的描述和着色做出富有创意的解读。这一变化从南华街和泰特美术馆的维多利亚式建筑中柔和的砖色到Blue Fin的钢架和玻璃中就能够体现出来。

冬景花园的铁锈红与泰特美术馆以及附近的Blackfriars桥的颜色遥相呼应，而外部的木材复合板和百叶窗则为大楼制造出一种温暖

旧场地地图，1682年 historic site map 1682

1800年 1800

1916年 1916

场地地图，2006年 site map 2006

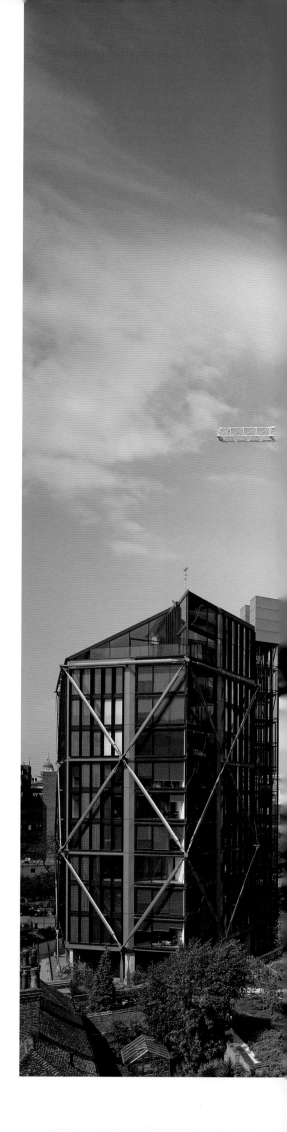

如家的感觉。凉亭独特的外部支撑系统已然移除了内部对承重墙的需求，同时也使得公寓之间的空间具有了高度的伸缩性。

这一支撑系统位于复合板外部，使得其自身可以被当作一个既特别又容易辨别的体系，从而也赋予了整个项目更高的魅力指数。它为建筑物立面增添了色彩和深度，而且还提供了一个可以将小号板材与大号板材进行统一的一种测算设施。有趣的是，许多买家纷纷要求购买窗外有节点的公寓，可见，带有支撑系统和节点的非同寻常的外表已然成为一个卖点。

带有玻璃的升降机塔为所有的居住者提供了一个纵览伦敦和泰晤士河以及每栋建筑东侧动感垂直交通流线的美好景致。冬景花园即是被围合起来的、每栋楼房北侧和南侧尽头都设有的单层玻璃阳台，悬于一个轻质的带有滑动屏风的平台的主要结构之外。它们可以被当作是封闭式露台，同时也算是屋内生活空间的一个外延。花园则成为整座建筑的"头部"，以钢质平台的形式呈现出来。

NEO Bankside

This scheme lies in the heart of the Bankside area of London, where a wide range of cultural, business and residential communities and a diverse and thriving built environment have co-existed in a web of streets dating from medieval times. The site is located close to the River Thames and is directly opposite the west entrance to the Tate Modern and the site of the proposed extension to the museum. The steel and glass pavilions fit perfectly into the Bankside landscape, reflecting the district's industrial past – and ushering into its exciting, residential future NEO Bankside comprises 217 residential units in buildings ranging from 12 to 24 stories. All the buildings of the scheme take their cues from the immediate context and it is the quality of the entire ensemble – rather than the individual parts – which creates drama. Landscaped groves define two clear public routes through the

1 场地边缘划分建筑区
1. site edges define building zones

2 公共空间的可渗透性有所提高，增加区域内的连接性
2. improved permeability for the public, improves connectivity within the area

3 街道形成了一个对角网格，从而明确了单座建筑物的轮廓
3. street creates a diagonal grid which then defines the profiles for the individual building

4 主建筑
4. principal buildings

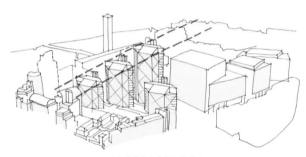

呈上升趋势的体块设计战略
rising massing strategy

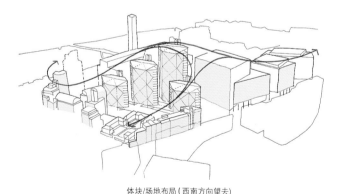

体块/场地布局（西南方向望去）
massing/site organization looking from southwest

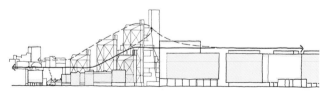

沿着南华街设置的与"肩部"高度相呼应的体块
massing responding to "shoulder" heights along Southwark Street

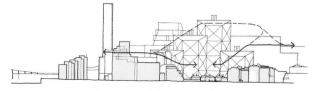

从霍普顿大街向东部望去，体块与公立救济院位于显著的位置
massing from Hopton Street looking east with Almshouses in foreground

163

site which extends the existing landscape from the riverside gardens outside Tate Modern through to Southwark Street.

NEO Bankside's four hexagonal pavilions have been imaginatively arranged to provide residents with generous accommodation and maximum daylight. The elegant and peaceful, landscaped gardens integrate seamlessly with Tate Modern and its surroundings, providing public access during the day as well as a secure, private environment for residents to enjoy. The stunning view from a three-bedroom apartment looking northeast includes St Paul's Cathedral and the City of London, the River Thames, Tate Modern and the Millennium Bridge, whilst from a southwest facing apartment there are views of the London Eye, "Big Ben" and the Houses of Parliament. The capital's essential business, education, shopping and entertainment destinations are all within easy reach.

The overall design hints at the former industrial heritage of the area during the 19th and 20th centuries, responding in a contemporary language which reinterprets the coloration and materials of the local architectural character. More specifically, the proposal seeks to achieve a contemporary architectural language which responds creatively to the articulation and coloration of the local context. This ranges from the warm brick hues of the Victorian buildings on Southwark Street and Tate Modern to the Blue Fin's precise steel and glass.

The oxide reds of the Winter Gardens echo those of Tate Modern and nearby Blackfriars Bridge, while the exterior's timber clad panels and window louvers give the building a warm, residential feeling. The pavilions' distinctive external bracing system has removed the need for internal structural walls and creates highly flexible spaces inside the apartments.

The bracing is located outside of the cladding plane allowing it to be expressed as the distinct and legible system which gives the scheme much of its charismatic language. The bracing gives a greater richness and depth to the facade and provides a scaling device which helps unify the micro scale of the cladding with the macro scale of the buildings. Interestingly, the dramatic appearance of the bracing and nodes has become a selling point, with many buyers requesting apartments with nodes outside their windows.

Glazed lift towers provide all occupants great views of London and the river, and a dynamic expression of the vertical circulation on the eastern side of each building. Winter gardens are enclosed, single-glazed balconies at the north and south ends of each building, are suspended from the main structure on a lightweight deck with large sliding screens. They act both as enclosed terraces and additions to the interior living space. The gardens effectively create "prows" and are expressed as exposed steel decks.

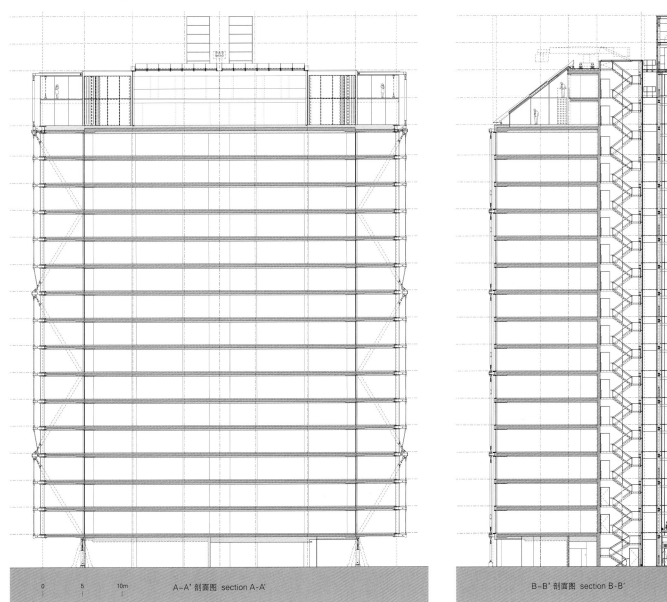

A-A' 剖面图 section A-A'

B-B' 剖面图 section B-B'

项目名称：NEO Bankside
地点：London, United Kingdom
建筑师：Rogers Stirk Harbour+Partners
合作建筑师：John Robertson Architects
项目经理：EC Harris
结构工程师：Waterman Structures Limited
服务工程师：Hoare Lea
景观设计师：Gillespies LLP
甲方：GC Bankside LLP (a Joint Venture between Native Land and Grosvenor)
用地面积：5,000m²
总面积：30,160m²
 -住宅区+办公区：28,600m²
 -零售区+地下室：1,560m²
造价：GBP 250m
竣工时间：2012
摄影师：
©Edmund Sumner - p.158~159, p.161, p.162, p.164, p.167
©Native Land (courtesy of the architect) - p.166 (except as noted)

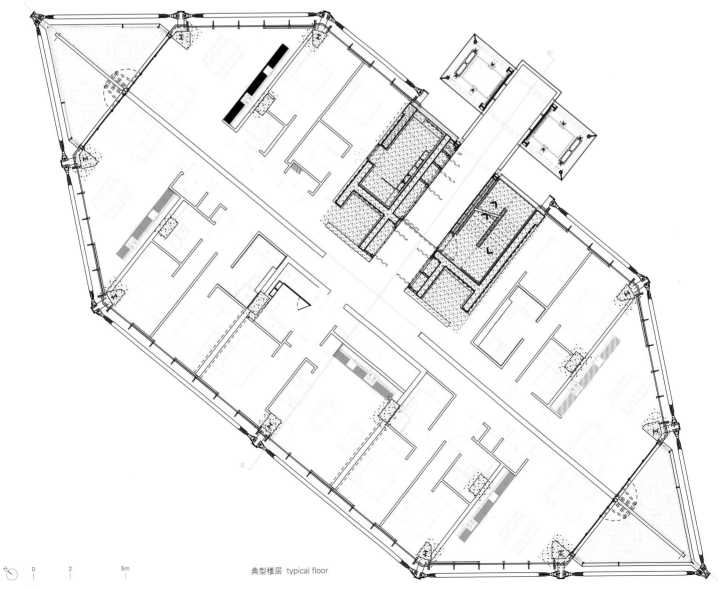

典型楼层 typical floor

来福士广场
Steven Holl Architects

凯德置地（中国）来福士城

"来福士城"品牌是"综合性建筑体"的代名词,由凯德置地(中国)开发、拥有、经营。"城"既代表着其所处的优越位置,也体现了多功能的"城中之城"这一理念。

"来福士城"品牌始于1986年的新加坡,现在在全球共有九个"来福士城"开发项目,其中八个在中国。上海来福士与北京来福士城分别于2003年和2009年投入运营,成都来福士广场购物中心和宁波来福士城也于2012年9月开张营业。其他的产业类型也会陆续投入使用。现在,位于杭州、深圳、上海(长宁区)和重庆的来福士城项目正在建设当中。

"来福士城"品牌开发项目取得的巨大成功具有里程碑意义,也是城市化的重要标志。这些开发项目都分别位于城市的交通枢纽附近,引来众多经济项目落户,同时也提高了人们的生活、工作以及购物水平。

来福士中国基金会向中国五个来福士开发项目投入了11.8亿美元,以提供强有力的资金支持,确保项目的稳定运营,促进资本提升。

项目

成都来福士广场位于成都市中心一环路和人民南路交汇处,"来福士广场"创造了巨大的公共广场,是多功能的城市综合体。项目旨在建造一处都市公共空间,而不是仅仅打造一个地标性摩天大楼。项目面积为278 000m²,建筑形状来源于自然光线的分布。为保证周边城市体块获得最小化的日光直射,城市肌理规定洞口高度为1.8m,满足地震对角线要求;而"被切开"的剖面覆有玻璃。

街区中心框定出的巨大的公共空间组成了三座"河谷",设计灵感为古代成都的伟大诗人杜甫(713—770)的名句"支离东北风尘际,漂泊西南天地间,三峡楼台淹日月,五溪衣服共云山"。三个广场上流水花园的设计以时间观念为基础,分别寓意着中国的年、月、日。三个水池充当下方六层高的购物中心的天窗。

建筑师应用"微型都市"的理念,在大都会场地内创造了适合人居住尺度的建筑,双门商店朝向街道和购物中心开放。塔楼体块上雕刻有三个巨大的洞口,分别是斯蒂芬·霍尔建筑师事务所设计的历史馆、利布斯·伍兹设计的光之馆和中国雕塑家韩美林设计的本土艺术馆。

成都来福士广场由468个地热井供热和制冷,广场中巨大的池塘可回收雨水,天然草地和睡莲带来了天然的冷却效果。项目采用高性能的玻璃、高效节能设备和当地材料等多种措施达到LEED黄金认证标准。

Sliced Porosity Block

About CapitaLand China Raffles City

The Raffles City brand is synonymous with a series of signature mixed developments, which are developed, owned and managed by CapitaLand China. The word "City" is representative of both the prime locations that the developments occupy as well as their multifunctional "city within a city" concept.

The Raffles City brand originated in Singapore in 1986 and now has a portfolio of nine Raffles City developments globally, with eight of them in China. Raffles City Shanghai and Raffles City Beijing were put into operation in 2003 and 2009 respectively. The shopping malls of Raffles City Chengdu and Raffles City Ningbo opened in September 2012 and other property types will come into service successively. Raffles City projects in Hangzhou, Shenzhen, Shanghai (Changning) and Chongqing are now under construction.

The highly successful Raffles City-branded developments are landmarks, as well as important symbols of urbanization. Located near the respective city's transportation hubs, these developments are economic magnets which serve to enhance people's living, working and shopping experiences.

The 1.18 billion USD Raffles City China Fund, injected in five raffles developments in China, provides strong financial support for the projects, ensures stable operation and promotes capital enhancement.

Project

In the center of Chengdu, China, at the intersection of the first Ring Road and Renmin Nam Road, the Sliced Porosity Block forms large public plazas with a hybrid of different functions. Creating a metropolitan public space instead of object-icon skyscrapers,

项目名称：Sliced Porosity Block – CapitaLand Raffles City Chengdu
地点：Chengdu, China
设计者：Steven Holl Architects
设计建筑师：Steven Holl, Li Hu
副主管：Roberto Bannura
项目建筑师：Lan Wu(Beijing), Haiko Cornelissen, Peter Englaender, JongSeo Lee(New York)
项目设计师：Christiane Deptolla, Inge Goudsmit, Jackie Luk, Maki Matsubayashi, Sarah Nichols, Manta Weihermann, Martin Zimmerli
项目团队：Justin Allen, Jason Anderson, Francesco Bartolozzi, Guanlan Cao, Yimei Chan, Sofie Holm Christensen, Esin Erez, Ayat Fadaifard, Mingcheng Fu, Forrest Fulton, Runar Halldorsson, M.Emran Hossain, Joseph Kan, Suping Li, Tz-Li Lin, Yan Liu, Daijiro Nakayama, Pietro Peyron, Roberto Requejo, Elena Rojas-Danielsen, Michael Rusch, Ida Sze, Filipe Taboada, Ebbie Wisecarver, Human Tieliu Wu, Jin-Ling Yu
合作建筑师：China Academy of Building Research
LEED 顾问：Ove Arup & Partners
用地面积：17,500m²　总建筑面积：310,000m²　有效楼层面积：195,000m²
施工时间：2008~2012　竣工时间：2012
摄影师：
Courtesy of the architect-p.176, p.184, p.185 bottom
©Shu He (courtesy of the architect)-p.168~169, p.170~171, p.173, p.177
©Iwan Baan (courtesy of the architect)-p.178~179, p.185 top

this three million-square-foot project takes its shape from its distribution of natural light. The required minimum sunlight that exposures to the surrounding urban fabric prescribe precises geometric angles that slice the exoskeletal concrete frame of the structure. The building structure is white concrete organized in six foot high openings with earthquake diagonals as required while the "sliced" sections are glass.

The large public space framed in the center of the block is formed into three valleys inspired by a poem of the city's greatest poet, Du Fu (713-770), who wrote, "From the northeast storm-tossed to the southwest, time has left stranded in Three Valleys." The three plaza levels feature water gardens based on concepts of time – the Fountain of the Chinese Calendar Year, Fountain of Twelve Months, and Fountain of Thirty Days. These three ponds function as skylights to the six-story shopping precinct below.

Establishing human scale in this metropolitan rectangle is achieved through the concept of "micro urbanism", with double-fronted shops open to the street as well as the shopping center. Three large openings are sculpted into the mass of the towers as the sites of the pavilion of history, designed by Steven Holl Architects, the Light Pavilion by Lebbeus Woods, and the Local Art Pavilion by Chinese sculptor Han Meilin.

The Sliced Porosity Block is heated and cooled with 468 geothermal wells and the large ponds in the plaza harvest recycled rainwater, when the natural grasses and lily pads create a natural cooling effect. High performance glazing, energy-efficient equipment and the use of regional materials are among the other methods employed to reach the LEED Gold rating.

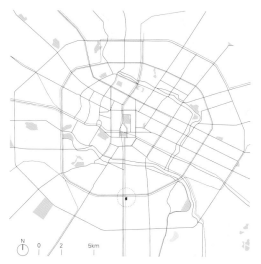

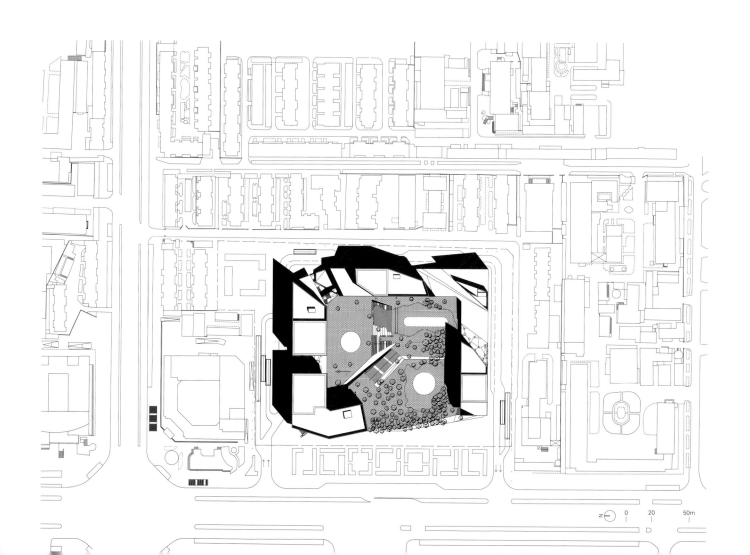

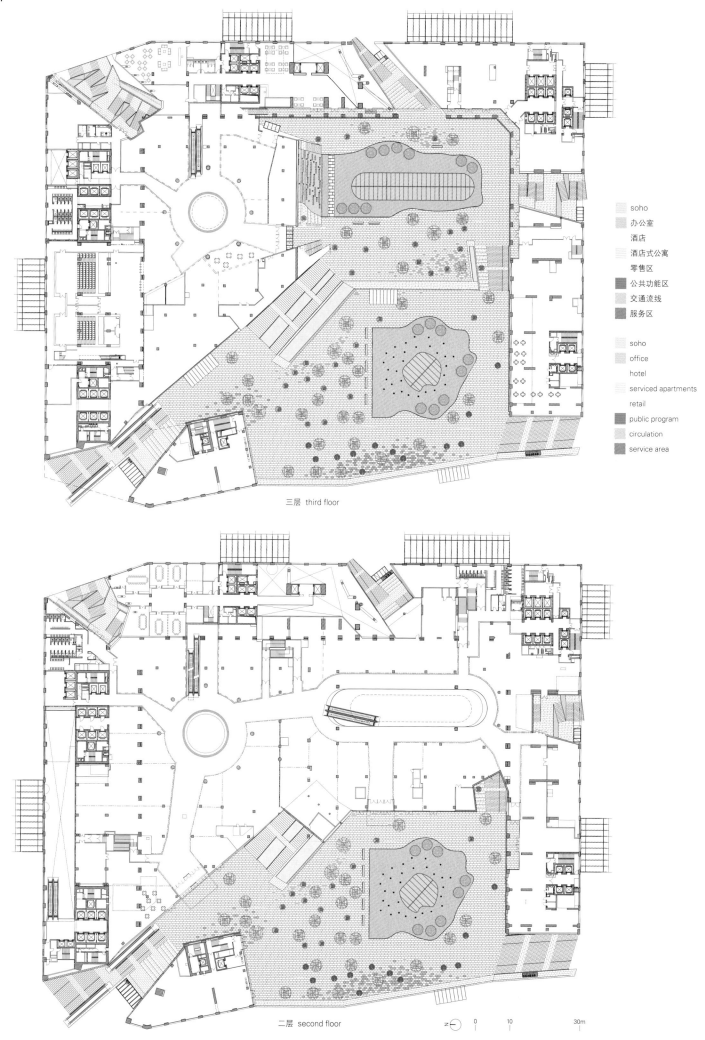

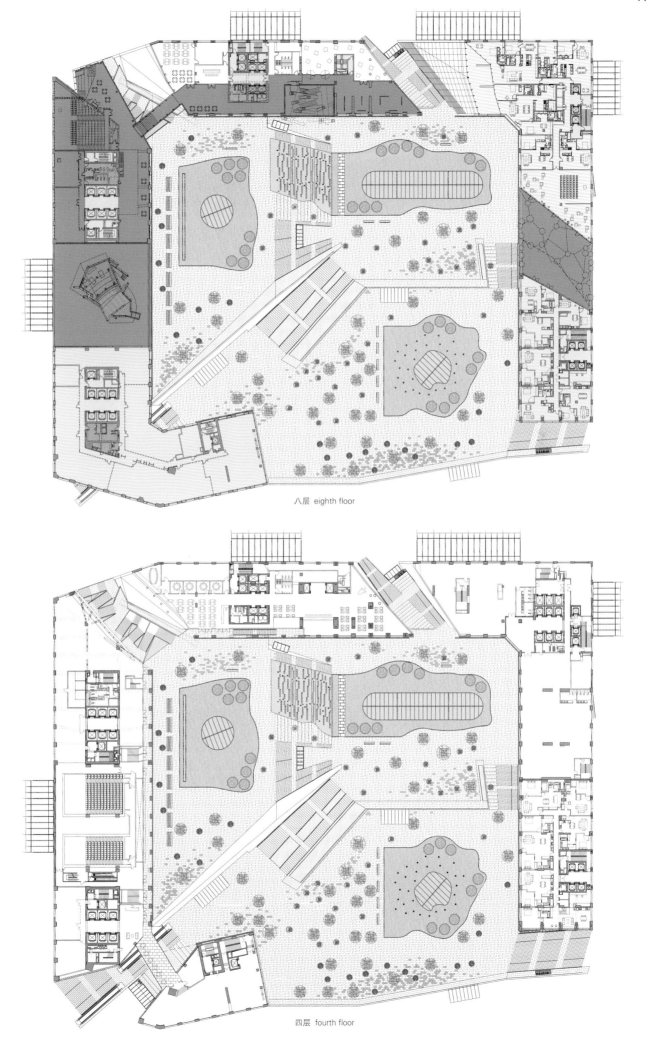

八层 eighth floor

四层 fourth floor

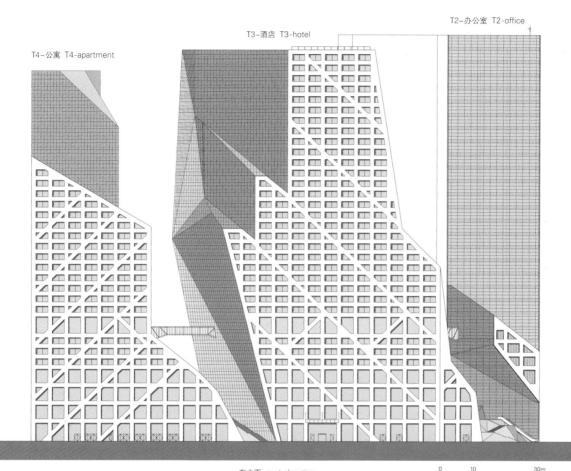

东立面 east elevation

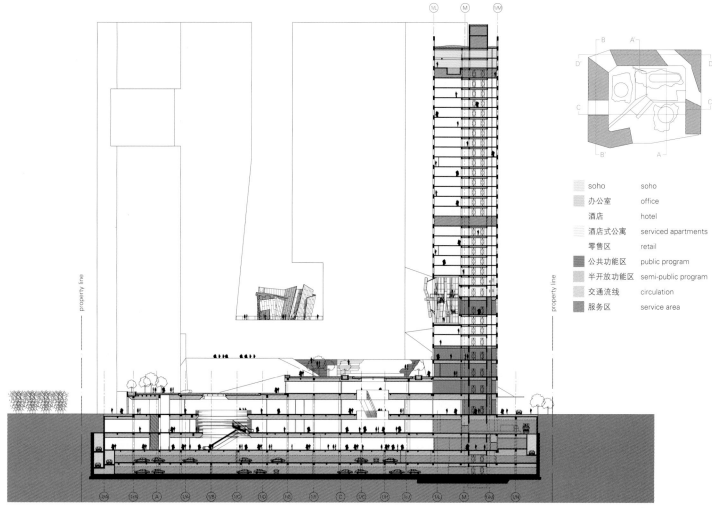

A-A' 剖面图 section A-A'

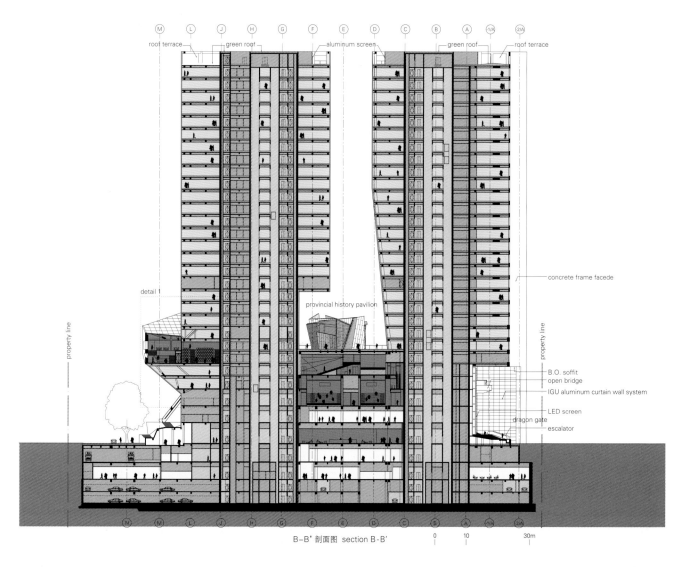

B-B' 剖面图 section B-B'

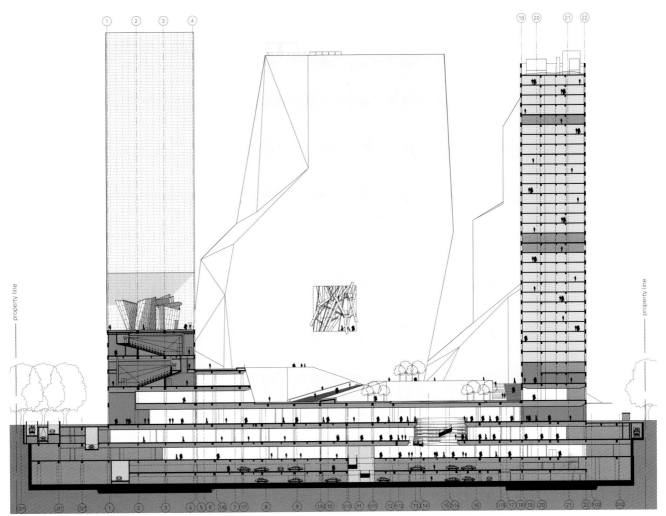

C-C'剖面图 section C-C'

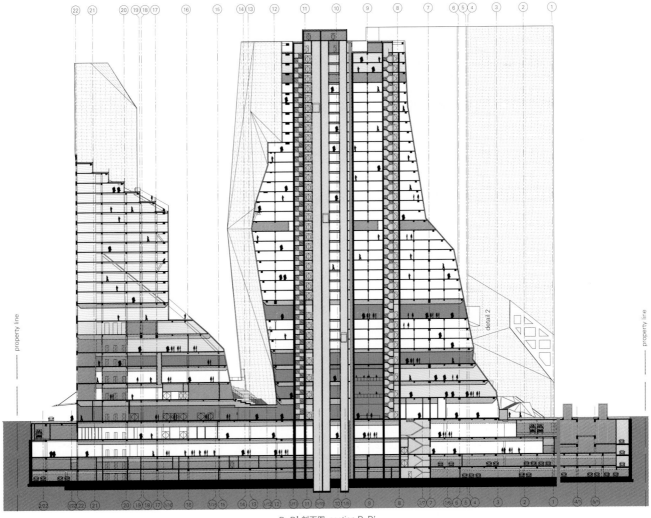

D-D'剖面图 section D-D'

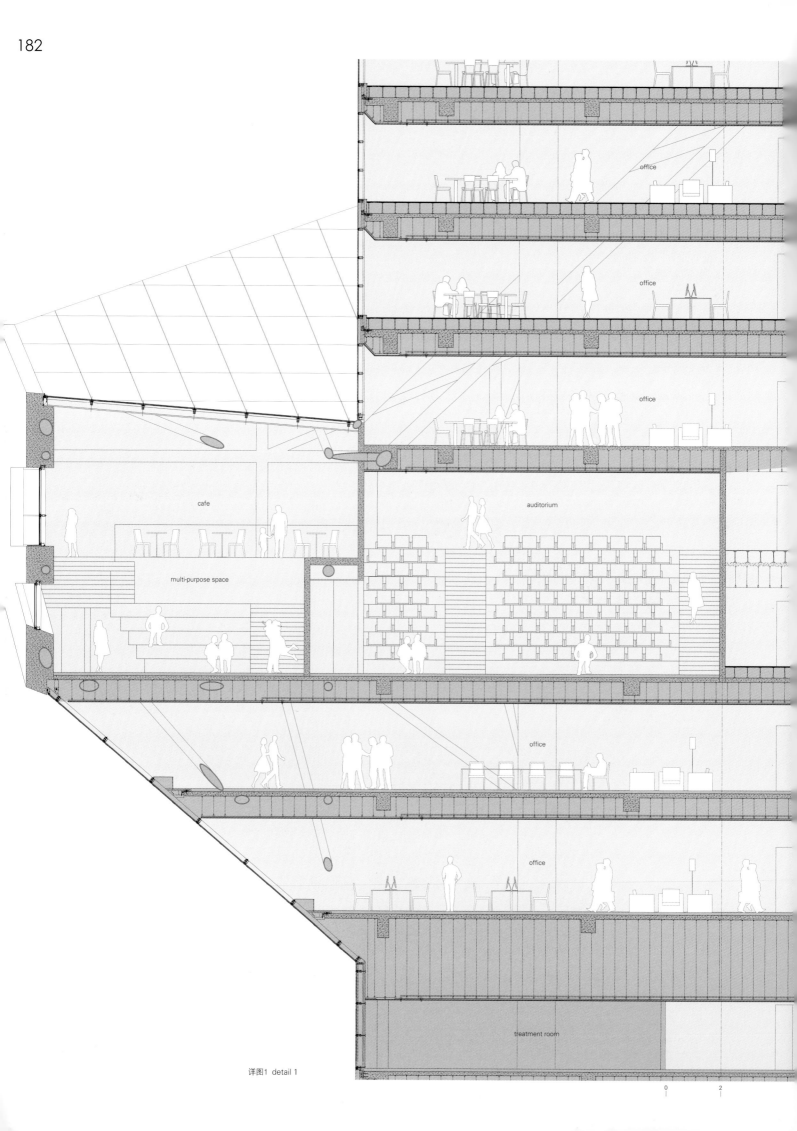

详图1 detail 1

1. building main structure
2. supply air duct shown in grey hatch
3. TYP. bridge roof assembly; waterproof membrane/light concrete screed/levelling/vapor barrier/insulation/metal deck
4. DIA 30mm painted steel tube handrail/bracket
5. bridge suspended ceiling system, plaster finished
6. hatchdoor for main tenance
7. insulated metal panel hatchdoor color match to roof mambrane
8. IGU TYP
9. steel tube beam building main structure
10. continuous ledge to slope along facade top of spandrel, shown in dashed
11. metal deck/light concrete infill
12. vapor barrier/thermal insulation/waterproofing
13. aluminum panel soffit, TYP.
14. soffit support substructure bolted to bridge main structure
15. bridge main structure
16. seismic joint @soffit and curtain wall connection

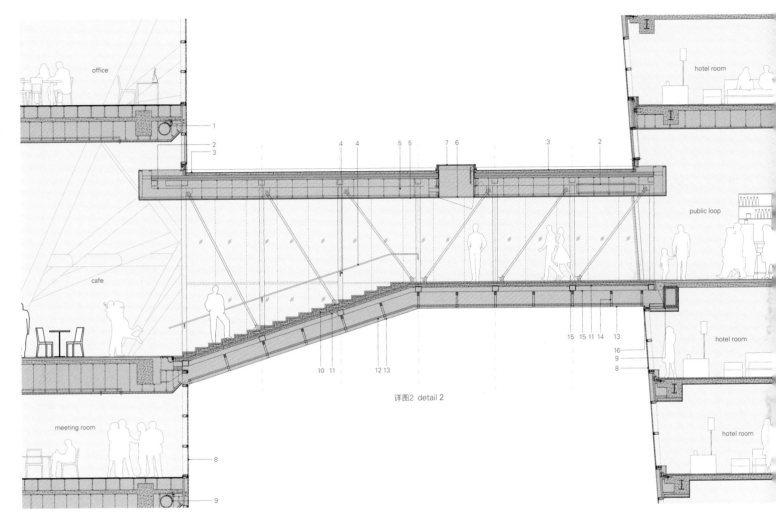

详图2 detail 2

>>92
SUGAWARADAISUKE
Daisuke Sugawara was born in Tokyo, Japan in 1977. Graduated from Architecture of Nihon University School in 2000 and received a master's degree from Graduate School of Science & Engineering, Waseda University in 2003. Worked for Shigeru Ban Architects in France from 2006 to 2007 and established international architects and art-directors office SUGAWARADAISUKE in 2008. Is currently working on different areas like urbanism, landscape, architecture, interior, graphic design and branding.

>>136
N Maeda Atelier
Norisada Maeda was born in 1960 in Tokyo, Japan. Studied in Kyoto University from 1980 to 1985. Worked for Taisei Corporation from 1985 to 1990. Has been working as principal of N Maeda Atelier since 1990. Is currently a visiting professor of Nihon University, Tokyo University of Science, Kokushikan University and Kyoto Seika University.

>>120
Oppenheim Architecture + Design
Is an architectural and planning firm based in Miami with offices in Los Angeles and Basel. Their specialties include the architecture, planning and interior design of large complex mixed-use waterfronts, hotels and resorts, retail and commercial offices, and residential buildings of all types, and award-winning work has been published several architectural publications such as The New York Times and Architectural Record. Chad Oppenheim established the firm in 1999, and has received a over 50 career distinctions including over 40 AIA Awards. Beat Huesler is the European partner of Chad Oppenheim who co-founded Basel office in 2009. Is a licensed architect with over 27 years of professional experience.

>>158
Rogers Strik Harbour + Partners
Is an international architectural practice based in London. Over 3 decades, RSHP has attracted critical acclaim and awards with built projects across Europe, North America and Asia. Is experienced in designing a wide range of building types including office, residential, transport, education, culture, leisure, retail, civic and health care. Graham Stirk was born in Leeds, 1957. Received 1st Class BA from Oxford Polytechnic. Since 1995, he has been a senior director of RSHP.

>>72
D·Lim Architects
Is an architectural group led by its two principals; YeongHwan Lim[left] and SunHyun Kim[right]. Lim is a professor of HongIk University and a registered architect in Pennsylvania. He is responsible for creative design. Kim holds a master's degree in project management from Harvard University and is responsible for the execution of projects. She is a registered architect in Korea. Both have received several architectural awards such as Seoul Architectural Award(2010), Young Architect Award(2010), "Best 7" Award from KIA(2011), Korean Architectural Award(2012). Their major projects are Ahn JungGeun Memorial, P&P Tower, She'smedi Hospital, H&M DongGyo, 1352 Lofts, and PHS CGOH Hospital.

Simone Corda
Is an architect and Ph.D candidate based in Sydney, Australia. Explores the themes of contemporary architecture through researches and projects at different scales and cross sectors. Referring to the architecture of Australia and New Zealand, he is currently focusing on the flexibility of housing as the key concept for sustainability. Part of his Ph.D thesis about Glenn Murcutt's work has been already published in the Italian magazine Area. Contributes to the Faculty of Architecture at the University of Cagliari enthusiastically with regular seminars and lectures at the Faculty of Architecture in Alghero, C.N.R. National Center of Research and Festarch event.

Paula Melâneo
Is an architect based in Lisbon. Graduated from the Lisbon Technical University in 1999 and received a master of science in Multimedia-Hypermedia from the cole Supérieure de Beaux-Arts de Paris in 2003. Besides the architecture practice, she focused on her professional activity in the editorial field, writing critics and articles specialized in architecture. Since 2001, she has been part of the editorial board of the portuguese magazine "arqa – Architecture and Art" and the editorial coordinator for the magazine since 2010. Has been a writer for several international magazines such as FRAME and AMC. Participated in the Architecture and Design Biennale EXD'11 as editor, part of the Experimentadesign team.

Alison Killing
Is an architect and urban designer based in Rotterdam, the Netherlands. Has written for several architecture and design magazines in the UK, contributing features and reviews to Blueprint and Icon and editing the research section of Middle East Art Design and Architecture. Most recently, she has worked as a correspondent for the online sustainability magazine Worldchanging. Has an eclectic design background, ranging from complex geometry and structural engineering, to humanitarian practice, to architecture and urban design and has worked internationally in the UK and the Netherlands, but also more widely in Europe, Switzerland, China and Russia.

>>44
Haworth Tompkins Architects
Was founded in 1991 by Graham Haworth[left] and Steve Tompkins[right]. Has an international reputation for award winning theater design. Was part of the Gold Award UK winning team at the Prague Quadrennial and was chosen to exhibit work at the 2012 Venice Biennale. Steve Tompkins is a contributor to Theatrum Mundi, a research forum led by Professor Richard Sennett, which brings architects and town planners together with performing and visual artists to re-imagine the public spaces of 21st century cities.

>>100
Abalo Alonso Arquitectos
Elizabeth Abalo[left] and Gonzalo Alonso[right] received a degree from the School of Architecture of the University of Navarra, Spain. Gonzalo Alonso was a teacher of projects at the University College Dublin from 2008 to 2011 and both of them at the University of A Coruña from 2011 to 2012. They have been working together since 1997 and obtained several awards and distinctions; among them the Coag Award 2012 for Rubido Romero Foundation, the Aplus Award 2011 for the Oleiros Health Center, the Spanish National Heritage Prize 2009 for the San Clemente Square or the galician Rodriguez Peña award 2007 for the San Pedro 78 House.

>>82
FGMF Arquitectos
Was established in 1999. Fernando Forte is an architect and urbanist graduated from FAU-USP, established FGMF while still in college. Specialized in structural systems and is in charge of coordinating constructions and monitoring implementations. Lourenço Gimenes received a master's degree in FAU-USP and also developed a Ph.D at the same institution. Is responsible for meeting partners and foreign consultants. Rogrigo Marcondes Ferraz graduated from FAU-USP. Currently develops a master's degree in architectural project. Acts as chief architect and complementary project teams coordinator.

>>168
Steven Holl Architects
Was founded in New York in 1976 and has offices in New York and Beijing. Steven Holl leads the office with partners Chris McVoy(New York) and Li Hu(Beijing). Graduated from the university of Washington and pursued architecture studies in Rome in 1970. Joined the Architectural Association in London in 1976. Is recognized for his ability to blend space and light with great contextual sensitivity and to utilize the unique qualities of each project to create a concept-driven design. Specializes in seamlessly integrating new projects into contexts with particular cultural and historic importance. Is a tenured faculty member at Columbia University where he taught since 1981.

>>30
Marie-José Van Hee
Was founded in 1975 and has been working closely together with Robbrecht en Daem Architecten on numerous projects since. Work of Van Hee is renewing the tradition of building timeless architecture with a particular attention to space, natural materials and light.

>>30
Robbrecht en Daem Architecten
Was founded in 1975. The main theme throughout the work is the relationship that they maintain between their architectural designs and the visual arts. Aims to establish a very contemporary and humane position, with their constructions making statements in a broad cultural context of science and art.

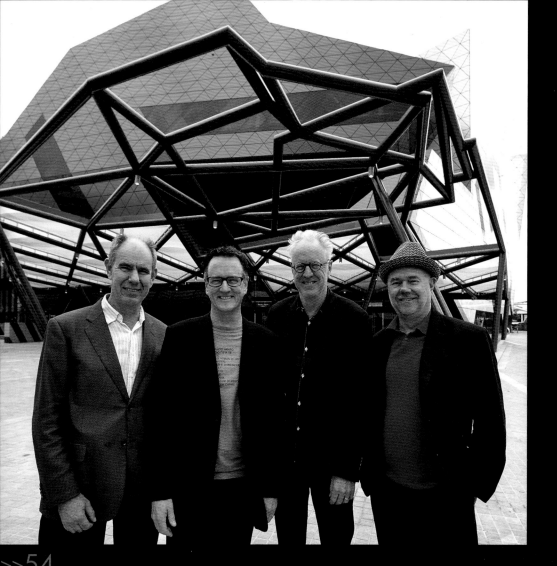

>>54

ARM Architecture
Was founded in 1988 by Stephen Ashton[first], Howard Raggatt[third], and Ian Mcdougall[fourth]. Tony Allen[second] is a managing director of ARM. ARM Architecture has been specialized in projects that require the highest level of design thinking and skillful resolution for over 20 years. From large urban master plans to exquisite and sensitive interventions, the hallmarks of an ARM Architecture project remain the same. Has received 40 state and national AIA awards and is the only firm in Australia to have been awarded the Victorian Architecture Medal four times.

>>148

Kiyonobu Nakagame Architects and Associates
Kiyonobu Nakagame was born in Kanagawa, Japan in 1965. Graduated from Nihon University School of Engineering department of Architecture in 1989. From 1990 to 1992, he worked for Shunji Kondo Architects. Established Kiyonobu Nakagame Architects and Associates in 1995 and taught students in Nihon University from 2007 to 2012.

>>110

MVN Arquitectos
Is an architectural firm formed in 2005 with the aim of improving in each project the individual's relationship with its environment through architecture. This principle is the argument of their design work, based on the use of light, energy and natural resources. Understands that this is the best way to adapt to the environment we live in a sustainable manner, valuing our environment as the main source of welfare, the cheapest and the most respectful. Diego Varela graduated from ETSAM in 1996. From 1997 to 2005, he worked for several architectural offices in Europe including Norman Foster & Partners and IDOM-ACXT, before establishing MVN Arquitectos. Emilio Medina also graduated from ETSAM in 1992. He worked for IDOM-ACXT from 2000 to 2004 and AVA Architects from 2004 to 2005. Co-founded MVN Arquitectos in 2005.

C3,Issue 2013.7

All Rights Reserved. Authorized translation from the Korean-English language edition published by C3 Publishing Co., Seoul.

© 2013大连理工大学出版社
著作权合同登记06-2013年第228号

版权所有·侵权必究

图书在版编目(CIP)数据

微工作·微空间：汉英对照 / 韩国C3出版公社编；王凤霞，杨薏，于风军译. — 大连：大连理工大学出版社，2013.10
（C3建筑立场系列丛书；32）
ISBN 978-7-5611-8255-0

Ⅰ．①微… Ⅱ．①韩… ②王… ③杨… ④于… Ⅲ．①办公建筑－建筑设计－汉、英 Ⅳ．①TU243

中国版本图书馆CIP数据核字(2013)第234394号

出版发行：大连理工大学出版社
　　　　　（地址：大连市软件园路80号　邮编：116023）
印　　刷：北京雅昌彩色印刷有限公司
幅面尺寸：225mm×300mm
印　　张：12
出版时间：2013年10月第1版
印刷时间：2013年10月第1次印刷
出 版 人：金英伟
统　　筹：房　磊
责任编辑：张昕焱
封面设计：王志峰
责任校对：高　文

书　　号：ISBN 978-7-5611-8255-0
定　　价：228.00元

发　行：0411-84708842
传　真：0411-84701466
E-mail：12282980@qq.com
URL：http://www.dutp.cn